# Excursions in number theory

# EXCURSIONS IN NUMBER THEORY

C. STANLEY OGILVY

*Professor Emeritus of Mathematics, Hamilton College*

JOHN T. ANDERSON

*Professor of Mathematics, Hamilton College*

DOVER PUBLICATIONS, INC.
*New York*

This Dover edition, first published in 1988, is an unabridged and slightly corrected republication of the work first published by Oxford University Press, New York, N.Y. in 1966. It is reprinted by special arrangement with Oxford University Press, 200 Madison Avenue, New York, N.Y. 10016.

Manufactured in the United States of America
Dover Publications, Inc., 31 East 2nd Street, Mineola, N.Y. 11501

**Library of Congress Cataloging-in-Publication Data**

Ogilvy, C. Stanley (Charles Stanley), 1913–
    Excursions in number theory / C. Stanley Ogilvy, John T. Anderson.
        p.    cm.
    Reprint. Originally published: New York : Oxford University Press, 1966.
    Includes index.
    ISBN 0-486-25778-9 (pbk.)
    1. Numbers, Theory of.   I. Anderson, John T.   II. Title.
QA241.035   1988
512′.7—dc19                                        88-20311
                                                        CIP

# Contents

# Excursions in number theory

# I
# The beginnings

In the beginning there were no numbers; or if there were, primitive man was unaware of them. Whether the numbers were always "there" (where?), or had to be invented, has been a much discussed question, and we shall leave it to the philosophers to continue that discussion without our aid. What we can say with some assurance is that the ability to count came relatively late to civilization. Nineteenth-century naturalists claimed that some animals could count up to 5. Early man could not do as well, and there are known to be isolated primitive tribes even today where any quantity more than 3 is known simply as "many."

Suppose we had only the quantities none, few, and many to count with. The addition table for school children would then be a simple one:

None + none = none
none + few = few          few + few = ?
none + many = many     few + many = many     many + many = many

The question mark indicates the only difficulty in the table. When does few plus few become not few but many? Such

quandaries led to the art of counting, the development of
which must have been a long and gradual process. Numbers
were not invented in an afternoon.

○

How does multiplication work under our present number
system? Let us dissect the multiplication of 307 by 43, by
examining instead the product of the two algebraic expressions

$$(3x^2 + 7) \times (4x + 3) = 12x^3 + 9x^2 + 28x + 21.$$

If now you substitute 10 for $x$ wherever it appears, you will
find that you have done the original multiplication of the
two numbers "by algebra." The reason it works is that 307
means

$$3 \times 10^2 + 0 \times 10^1 + 7 \times 10^0,$$

where $10^0$ means 1.  (Why?)

Note the importance of zero as a spacer, or place-holder;
307 is not the same as 37, although $3x^2 + 0 + 7$ *is* the same
as $3x^2 + 7$. Here something else is acting as the place-
indicator. (What?) Thus our number system is a shorthand
for *the algebra of polynomials in one variable* with the variable
replaced by the number 10.

What is so sacred about 10? It was chosen, in all likelihood, because people had ten fingers. The algebra would be
correct whether the variable were 10 or some other number
or just $x$. Could 12 or 7 have been used instead of 10?
Certainly, and if you are under eighteen you are much more
likely than are your parents to be acquainted with arithmetic
in bases other than 10. It is a topic now covered in many
elementary and secondary schools as part of the "new

mathematics." It is not really very new; but is has come into its own, so to speak, as a result of the invention of the binary digital computer.

A modern electronic computer, which according to the popular notion is so clever that it can do almost anything, is in fact a sub-moron. What it can do is exceedingly fundamental: it can answer certain simple questions, like whether a given number is greater than another given number. It can only answer this question with "yes" or "no"; but it can do it rapidly, and then proceed to the next question and answer it correctly, and so on, so that it builds up (through the instructions of the man operating the machine) a complicated sequence of operations in a time span unbelievably shorter than that required by a human being for the same sequence. The computer does all this by the simple process of accumulating yes's and no's in the correct combinations. Each yes is accomplished by an "on" signal and each no by an "off" signal. Therefore all the numerical material has to be translated to an "on-off" system before it can be fed into the machine. For this purpose one uses the *binary* system, whose base is not 10 but 2. It is the simplest system available in that it requires only the digits 0 and 1. (Why not 2?) Thus 0 can be represented by "off" and 1 by "on."

The binary system, though ideal for the computer, would be awkward for everyday use. What does our algebra look like if we replace $x$ by 2? We have then, for $3x^2 + 7$, $3 \times 2^2 + 7$. But already we are in trouble, because 3 and 7 are not supposed to be in the system. 7 itself would have to be written $2^2 + 2^1 + 2^0$; or in algebra, $x^2 + x + 1$, where $x$ is now 2. In binary notation this is the number 111, meaning not one hundred and eleven, but one two-squared plus one two-to-the-first-power plus one. The number 307 comes out, in binary notation, to be 100110011, which would not have the convenience of 307 at the grocery store, perhaps, but is duck soup for the computer.

Until recent years the binary system was looked upon as something of a mathematical curio of only theoretical importance. Suddenly it has become indispensable, and would have had to be developed in a hurry had it not been ready to hand. This has been the story again and again throughout the history of mathematics: no part of mathematics is ever, in the long run, "useless." Most of number theory has very few "practical" applications. That does not reduce its importance, and if anything it enhances its fascination. No one can predict when what seems to be a most obscure theorem may suddenly be called upon to play some vital and hitherto unsuspected role.

○

Write down the number 22.  Divide by 2, and          22   0
write the result underneath.  Divide by 2 again:     11   1
half of 11 is 5½, so throw away the ½ and write       5   1
simply 5.  Repeat the process, always discarding      2   0
the ½ when the division does not come out even.       1   1
Stop when you reach 1.

Now opposite each number in your column, write a zero if the number is even and a 1 if it is odd.  The resulting column, *read from the bottom up*, is the binary representation of 22: 10110.

As a test of your own mathematical aptitude you might try analyzing this scheme, or *algorithm*, to explain to yourself why it works, before reading further.

There is no mystery if we attack the problem correctly. As with so many mathematical problems, the question cannot be answered until it is properly asked. Many a worker, not only in mathematics but in other fields of research, has bogged down because he is asking himself the wrong question—probably one that has no answer. When somebody else comes along and rephrases the question or perhaps asks a new one, a breakthrough results.

The instructions have been presented in a way that conceals the inner workings of our little algorithm. Let us begin with a simpler example, one that we have already looked at: $7 = 2^2 + 2^1 + 2^0$, or 111 in the binary notation. Starting at the *right*-hand end, what does the *last* 1 mean? It means that 1 is contained in 7 an odd number of times, and hence there must be a 1 at the end because it cannot be absorbed in any other power of 2. Having written down that 1 as the last digit, we now have not 7 but 6 left to represent. Next we look at the next to last digit. It is a 1, which should mean that there is an odd number of 2's in 6. Correct! So we subtract 2, representing the fact that we are writing an abbreviation for $1 \times 2^1$, and move once more to the left asking is 4 contained in 4 an odd or an even number of times? 1 is odd, so 1 is written in the $2^2$ slot and the representation is complete.

On the other hand suppose we had started with 9. The 1's digit is the same, namely, 1, so we subtract 1 from 9 and examine the 8. But 2 goes an *even* number of times into 8. Hence we do *not* want a 1 in the $2^1$ place, because an even number of 2's can be absorbed in the 4's place. So we write a 0 and move to the left, still operating on 8 because 0 takes nothing away from 8. Now 4 also divides 8 an even number of times, so we write another 0 and move left again. Eight divides 8 once, for a net result of 1001 for 9.

Now how about 22? Forget the language in which the algorithm was phrased, especially the part about discarding

$\frac{1}{2}$'s. Look at 22 and ask is 1 contained in it an odd or an
even number of times? Even; so write a 0 for the last digit
of the binary answer. Next, is 2 contained in 22 an odd or
an even number of times? Odd; so write a 1. But, in the
2's place this represents $2^1$, which therefore has been "used"
and must be subtracted from 22, leaving 20. At the next
stage, 4 is contained in 20 an odd number of times (5), so a
1 should appear opposite the 5 (it is never a question of
"throwing away" $\frac{1}{2}$). Now because $1 \times 2^2 = 4$, the next
question is how many 8's in 16, not 20. Two (even), and
hence a 0 is written. Finally, how many 16's in 16? One,
so 1 is the bottom digit. What we have discovered is that

$$22 = 1 \times 16 + 0 \times 8 + 1 \times 4 + 1 \times 2 + 0 \times 1,$$

or 10110.

We have done this example in such detail in order to
explain a form of multiplication actually in use (at least
until recently) in some remote corners of the world. The
following instance came to our attention a few years ago.
A certain colonel once made an expedition into Ethiopia.
Somewhere in the far interior his party had occasion to
purchase seven bulls.

"This we attempted to do," he writes, "at the first market
place we came to, but although there were bulls for sale
there, neither the owner of the stock nor my headsman knew
how many Maria Theresa dollars should change hands.
As neither could do simple arithmetic, they just stood and
yelled at each other, getting nowhere. Finally a call was
put in for the local priest, as he was the only one who could
handle questions like this.

"The priest and his boy helper arrived and began to dig
a series of holes in the ground, each about the size of a

teacup. These holes were ranged in two parallel columns; my interpreter said they were called houses. What they were about to do covered the entire range of mathematics necessary to transact business in this area, and the only requirement was the ability to count, and to multiply and divide by two.

"The priest's boy had a bag full of little pebbles. Into the first cup of the first column he put seven stones (one for each bull), and 22 into the first cup of the second column, since each bull was to cost $MT 22. It was explained to me that the first column is used for multiplying by two: that is, twice the number of pebbles in the first house are placed in the second, then twice that number in the third, and so on. The second column is for dividing by 2: half the number of pebbles in the first cup are placed in the second, and so on down until there is one pebble in the last cup. Fractions are discarded.

"The division column is then examined for odd or even number of pebbles in the cups. All even houses are considered to be evil ones, all odd houses good. Whenever an evil house is discovered, the pebbles are thrown out and not counted. All pebbles left in the remaining cups of the multiplication column are then counted, and the total of them is the answer."

On paper, the problem of the bulls looks like this:

| Multiplication Column | Division Column |
|:---:|:---:|
| 7 | 22 |
| 14 | 11 |
| 28 | 5 |
| 56 | 2 |
| 112 | 1 |
| 154 | |

The colonel marvels at the fact that the correct result is always obtained by this system of multiplication, even though half of five is not two (he calls this an error). But we know there is no error. If we factor a 7 out of all remaining members of the left-hand column (leaving 2, 4, and 16) and recall how we just obtained 10110 for 22, it is clear that the total of the left-hand column represents $7 \times 22 = 154$.

○

There is an old familiar game for two players, called Nim. From three piles, each pile containing a random number of chips, the two players take turns in removing one or more chips (or the whole pile) from any *one* pile. The object of the game is to force the opponent into the position where he must remove the last chip.

There is a "best way" to play this game, and it is sure to win against a player who does not understand it. If both players know the best strategy, the one who is allowed the first move is heavily favored. The analysis depends on the binary system. If we were accustomed to counting with the base 2 the game would be a simple one. In our explanation (see Notes, p. 146), we do not actually change to binary notation; we merely arrange the counters or chips in powers of 2, which amounts to the same thing.

○

In the Roman numeral system a number is recorded as a series of hatch marks, with special symbols for certain numbers like 5, 10, 50, 100, and 500. 5 is V, 50 is L, and so forth—but without rhyme or reason. The special letters are simply *assigned*.

The importance of zero lies not so much in its existence as in its use. In the system of Roman numerals there are no zeros. V and L are not related symbols, whereas in our (Arabic) system 5 and 50 are related, and in fact are distinguishable only because of the position of the zero. Zero, then, becomes important as a place holder.

We have previously indicated that the ease of multiplication in our system of numerals to the base ten depends on the algebraic properties of that system. No such properties can be assigned to Roman numerals. Several methods by which the Romans might have multiplied two large numbers have been suggested, but all are complicated and difficult. Try it! The chances are that most Romans, if they ever multiplied at all, carried out a repeated addition. Roman addition presents no serious difficulties: in fact it is easier than ours. First one changes all numbers into the form in which no subtractions are indicated. For instance 49, normally written XLIX, would be changed to XXXXVIIII for purposes of addition. After that, the only thing to do is list letters in suitable columns and be careful about carrying:

|   | XXXXVIIII |    | 49  |
|---|-----------|----|-----|
| CL | X | II | 162 |
| CC | X | I  | 211 |

# 2

# Number patterns

Given the number 123456789, in how many ways can the digits (numerals) of this number be rearranged to form new numbers; and how many of them are divisible by 3— meaning evenly divisible, with remainder zero?

The answers to both these questions are very simple. Including the present arrangement, there are 9 ways of choosing the first digit; then, when that has been done, there are 8 ways of choosing the second digit; then 7 for the third, and so on, for a total of

$$9 \times 8 \times 7 \times 6 \times 5 \times 4 \times 3 \times 2 \times 1 = 362{,}880$$

different numbers. And how many are divisible by 3? All of them.

A number is divisible by 3 if and only if the sum of its digits is divisible by 3. Since

$$1 + 2 + 3 + 4 + 5 + 6 + 7 + 8 + 9 = 45,$$

and since 45 is divisible by 3, so is the original number, and of course permuting the order of the digits does not alter their sum.

We might note in passing an easy way to sum the first nine (or any other number) of consecutive integers. They can be grouped in equal pairs, working from both ends of the string:

$$1 + 9 = 10$$
$$2 + 8 = 10$$
$$3 + 7 = 10$$
$$4 + 6 = 10$$
$$5 = \underline{\ 5\ } \quad \text{(no mate, because the collection contained an } \textit{odd} \text{ number of objects)}$$

Total 45

Carl Friedrich Gauss, possibly the greatest mathematician of all time, showed his arithmetical skill at an early age. When he was ten years old his class at school was given what was intended to be a long routine drill exercise by a tyrannical schoolmaster: "Find the sum of the first 100 positive integers." This was easy for the schoolmaster, who knew how to sum arithmetic progressions, but the formula was unknown to the boys. Young Gauss did not know how to do it either, but he invented a way, instantly and in his head. Writing the answer on his slate, he handed it in at once. When the rest of the students' calculations were collected an hour later, all were found to be incorrect except Gauss's! We are told that he did it by pairing the terms and then mentally multiplying the value of each pair by the number of pairs. If the pairs could each total 100, so much the easier: $100 + 0$, $99 + 1$, etc. This would make 50 pairs of 100 each for 5000, plus 50 left over (the middle number), for a total of 5050.

Why does the old familiar test for division by 3 work the way it does? The test for divisibility by 9 is similar, and we can scoop them both up in the same basket. Let us use a number whose four digits are *abcd*. We have seen that this means

$$1000a + 100b + 10c + d.$$

Hence this number can be written as the sum of two numbers:

(1)  $999a + 99b + 9c$

and

(2)  $a + b + c + d.$

The number represented by (1) is certainly always divisible by 3 (or 9), regardless of the values of $a$, $b$, $c$, and $d$, and therefore (1) + (2) is divisible by 3 or (9) exactly when and only when (2) is so divisible. It can be seen that a number of five of more digits is handled in just the same way.

The test for divisibility by 11 is only slightly more complicated. We write our number in the form

$$1001\,a + 99\,b + 11\,c$$
$$-a + b - c + d.$$

Now the original number is divisible by 11 if and only if the quantity $(-a + b - c + d)$ is divisible by 11. The quotient may be negative or zero as long as it is a whole number. It remains to convince yourself that 9999, 100001, and so on, are all divisible by 11, which we leave to you to do by a brief study of the divisions in question. It is one of those things that is so "easy to see" that it is only clouded when encumbered by the wording of a formal proof.

○

In the natural sciences one puts great faith in *induction*, or reasoning from the particular to the general. In mathematics we cannot rely on any such process.

Someone says he has a formula for the sum of the first $n$ even numbers:

$$2 + 4 + 6 + \cdots + 2n.$$

He claims that this sum is always equal to

$$Y(n) = n^4 - 10n^3 + 36n^2 - 49n + 24.$$

The symbol $Y(n)$, read "$Y$ of $n$," means that if you put some particular $n$ into the right-hand side, the quantity $Y(n)$ will be the result. By trial, one finds that

$$Y(1) = 1 - 10 + 36 - 49 + 24 = 2.$$

Likewise $Y(2) = 6$, $Y(3) = 12$, and $Y(4) = 20$. The formula seems to be checking out. For

$$
\begin{aligned}
2 &= 2 \\
2 + 4 &= 6 \\
2 + 4 + 6 &= 12 \\
2 + 4 + 6 + 8 &= 20.
\end{aligned}
$$

By this time we are getting tired of so much arithmetic. Can we conclude that because the formula has worked for four cases it must be the correct one? We cannot. Indeed, $Y(5)$, which ought to be 30, is in fact 54; and the "formula" is never again correct for any other $n$.

There is a well-known formula that delivers *prime* numbers (see Notes) for the first 79 cases; but when yout try $n = 80$, the formula fails. The physicist would be content to risk a theory with far fewer than 79 experimental verifications. In mathematics neither seventy-nine nor a million and seventy-nine are enough. We must have a different sort of proof.

An important kind of reasoning called *mathematical* induction provides a very useful method of proof. Before we say what it is we had best be quite sure we understand what it is not.

Mathematical induction is not, as some people wrongly suppose it to be, examining a sequence and then reasoning by inference. Look at the following sequence:

$$
\begin{aligned}
1 &= 1 \\
1 + 3 &= 4 \\
1 + 3 + 5 &= 9 \\
1 + 3 + 5 + 7 &= 16 \\
1 + 3 + 5 + 7 + 9 &= 25.
\end{aligned}
$$

So far, the right-hand column turns out to be the sequence of the perfect squares. Does this assure us that the process will continue indefinitely? It certainly does not. The apparent law of formation is not a law until it is proven so, no matter how strong appearances may be.

What we should like to prove is that the first $n$ odd consecutive positive integers will always add up to $n^2$. This we shall do by mathematical induction—a very different thing from looking and guessing. The process can be likened to that of teaching a blind child how to climb a ladder. Whether it is ever done this way is of no importance: it *could* be done this way. First the child could be placed on some rung of the ladder, no matter which rung, and instructed how to get from there to the next higher rung. That done, the only remaining thing would be to make sure that the child knew how to find and climb onto the bottom rung. The child would thereafter know how to get to the second rung, and thence to the third, and so on.

We begin by making what is known as the induction assumption. We do not assume what we are trying to prove in general. We make the milder, less sweeping assumption that, for the time being, what we are trying to prove is true for a particular case, say $n = k$. This is equivalent to being placed on the $k$'th rung of the ladder. Well then, if the theorem about the odd integers were true for $n = k$, we would have

$$1 + 3 + 5 + \cdots + (2k - 1) = k^2.$$

If this were so, we could add $2k + 1$ (the next term) to both sides:

$$
\begin{aligned}
1 + 3 + 5 + \cdots + (2k - 1) + (2k + 1) \\
= k^2 + (2k + 1) \\
= k^2 + 2k + 1 \\
= (k + 1)^2
\end{aligned}
$$

This says that *if* the theorem is true for $n = k$ then it must be true *also* for $n = k + 1$, and we are now on the $(k +1)$st rung of the ladder. It does not yet say that it actually *is* true for any $n$ at all. But now we go back and look at some small value of $n$, usually the first, $n = 1$. The theorem is surely correct for that value: $1 = 1^2$ (we are on the lowest rung). Therefore, by our induction assumption and what it led to, it is true for $1 + 3$ (we are on the way up), and therefore for $1 + 3 + 5$, and so on indefinitely, and we have proved that, for any $n$,

$$1 + 3 + 5 + \cdots + (2n - 1) = n^2.$$

By setting $n = m^p$, one sees that $m^{2p}$, an even power of any integer, is equal to the sum of all the odd integers up to and including $2m^p - 1$; if $m = 3$ and $p = 2$, for example,

$$81 = 3^4 = 1 + 3 + 5 + \cdots + 17.$$

It can also be shown that any integral power of a whole number, say $m^k$ where $k > 1$, is the sum of a certain "block" of the odd integer sequence exactly $m$ units long. For instance, $m^3$ is always the sum of $m$ odd integers beginning with $m^2 - m + 1$ and ending with $m^2 + m - 1$. Thus if $m = 5$ we have the five odd integers starting at $5^2 - 5 + 1 = 21$ and ending at $5^2 + 5 - 1 = 29$:

$$5^3 = 21 + 23 + 25 + 27 + 29.$$

The method of proof by mathematical induction is exceedingly useful throughout virtually all branches of mathematics. It has one disadvantage: it does not *construct* a theorem from scratch. But if we have some hint, some intelligent guess, as to a plausible statement, we can often test its validity by the induction method.

○

The ancients used to think of the square numbers geometrically. They spoke of the square *on* the side of a triangle instead of the square *of* the side. This usage has come down to us in our common exponential terminology. We normally speak of the *square* of five rather than the *second power* of five even though we simply mean the number 25. The Greeks preferred to think of the picture shown in Figure 1. There are 25 square units in a square of side 5.

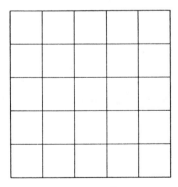

FIGURE 1

The next step was to think of the square numbers as square arrays of points:

$$
\begin{array}{cccc}
& & & \bullet \;\; \bullet \;\; \bullet \;\; \bullet \\
& & \bullet \;\; \bullet \;\; \bullet & \bullet \;\; \bullet \;\; \bullet \;\; \bullet \\
& \bullet \;\; \bullet & \bullet \;\; \bullet \;\; \bullet & \bullet \;\; \bullet \;\; \bullet \;\; \bullet \\
\bullet & \bullet \;\; \bullet & \bullet \;\; \bullet \;\; \bullet & \bullet \;\; \bullet \;\; \bullet \;\; \bullet \\
1^2 & 2^2 & 3^2 & 4^2
\end{array}
$$

The "triangular numbers" also received their share of attention:

$$
\begin{array}{cccc}
& & & \bullet \\
& & \bullet & \bullet \;\; \bullet \\
& \bullet & \bullet \;\; \bullet & \bullet \;\; \bullet \;\; \bullet \\
\bullet & \bullet \;\; \bullet & \bullet \;\; \bullet \;\; \bullet & \bullet \;\; \bullet \;\; \bullet \;\; \bullet \\
1 & 3 & 6 & 10
\end{array}
$$

A connection between the triangles and the squares is that the sum of any two consecutive triangular numbers is the square number whose "side" is the same as the side of the larger of the two triangles. $3 + 6 = 3^2$, $6 + 10 = 4^2$, etc. This is easy to prove by algebra; but it is also evident geometrically. Arrange the two triangles, say 6 and 10, thus:

$$
\begin{array}{cc}
\bullet & \bullet \\
\bullet \;\; \bullet & \bullet \;\; \bullet \\
\bullet \;\; \bullet \;\; \bullet & \bullet \;\; \bullet \;\; \bullet \\
& \bullet \;\; \bullet \;\; \bullet \;\; \bullet
\end{array}
$$

The first can always be fitted (upside down) upon the second:

$$
\begin{array}{c}
\bullet \;\; \bullet \;\; \bullet \;\; \bullet \\
\bullet \;\; \bullet \;\; \bullet \;\; \bullet \\
\bullet \;\; \bullet \;\; \bullet \;\; \bullet \\
\bullet \;\; \bullet \;\; \bullet \;\; \bullet
\end{array}
$$

The sum of the first $n$ positive integers is

$$1 + 2 + 3 + \cdots + n = \frac{n^2 + n}{2}.$$

From the way triangular numbers are formed, it is clear that this is also the formula for the $n$'th triangular number. The sum of the first $n$ squares likewise has an easy expression in terms of $n$:

$$1^2 + 2^2 + 3^2 + \cdots + n^2 = \frac{n(n + 1)(2n + 1)}{6}.$$

We leave it to you to prove these formulas by induction. Without knowing them ahead of time, it would take quite a bit of ingenuity to *derive* and prove them. But there are methods in the calculus of finite differences that solve such problems with routine ease.

As for the cubes, it turns out that

$$1^3 + 2^3 + 3^3 + \cdots + n^3 = \left(\frac{n^2 + n}{2}\right)^2.$$

Thus we have the unexpected relation that the sum of the first $n$ *cubes* is equal to the *square* of the sum of the first $n$ integers.

○

Can a square also be a triangular number? Certainly the number 1 is both square and triangular. Are there more? An infinite number? Examining the sequence of squares we find that 36 is the next one that is also triangular. But if we simply hunted among the squares the search might be a very long one. The next three squares that are also triangular are 1225, 41616, and 1413721. How were these found?

To answer this question we need more powerful equipment than we now have at our command. We shall return to the problem and solve it in Chapter 10.

○

To the ancients, whose science and mathematics were immersed in and confused by philosophy and metaphysics, numbers had personalities, and geometric figures were often endowed with pseudo-sacred qualities. Even as late as 1596 we find Johannes Kepler, one of the fathers of modern astronomy, defending a plan of the solar system on the grounds that the six planets (all that were then known) fitted into positions that depended on relations among the five regular geometric solids. It is not surprising, then, to find the name *perfect* applied to a number equal to the sum of all its proper divisors (including 1). Six is a perfect number being equal to $1 + 2 + 3$. The next is $28 = 1 + 2 + 4 + 7 + 14$. Noting that only 8 perfect numbers were (then) known, Mersenne wrote in 1644, "We see clearly from this fact how rare are the Perfect Numbers and how right we are to compare them with perfect men."

All known perfect numbers are even. It is not known whether any odd perfect numbers exist; but if there is one, it is certainly greater than $10^{25}$. A very large amount of time has been devoted to the study of perfect numbers. One prominent contemporary number theorist has used it successfully as the springboard for an entire book.

The only perfects known to the Greeks were

$$P_1 = 6 = 2(2^2 - 1)$$
$$P_2 = 28 = 2^2(2^3 - 1)$$
$$P_3 = 496 = 2^4(2^5 - 1)$$
$$P_4 = 8128 = 2^6(2^7 - 1).$$

From this scanty evidence it was conjectured during the Middle Ages (1) that there was probably a perfect number between each power of 10 and that therefore the $n$'th one was always $n$ digits long; and (2) that the perfect numbers end alternately in 6 and 8. Here we have another example of "intuitive induction" based on nothing but a wild guess. Both conjectures are wrong. There is no perfect number of 5 digits; in fact

$$P_5 = 2^{12}(2^{13} - 1) = 33550336$$

and while that does end in 6, the next one ends in 6 also, not 8. It *is* true that they always end in 6 or 8, but not alternately.

You will have noticed that we have made a point of displaying each of the perfect numbers in the form

$$2^{p-1}(2^p - 1).$$

Furthermore $p$ has been a prime in every case; yet after using $p = 2$, 3, 5, 7, we skipped 11 and said that the next perfect number was the one for which $p = 13$. All these facts play significant roles in the problem of the perfect numbers.

First let us prove what was known to Euclid: *If* $2^p - 1$ is prime, *then* $N = 2^{p-1}(2^p - 1)$ is perfect. To do this we list all the possible factors of $N$. Certainly each of 1, 2, $2^2$, ... $2^{p-1}$ is a factor. It is a hypothesis of the theorem that the parenthesis is prime, but as many more factors again can be obtained by multiplying the parenthesis by each factor in the list of powers of 2. This exhausts the possibilities: $N$ has no further factors. We have then two sets to add:

$$S_1 = 1 + 2 + 2^2 + \cdots + 2^{p-1}$$
$$S_2 = (1 + 2 + 2^2 + \cdots + 2^{p-1})(2^p - 1) = S_1(2^p - 1)$$
$$\therefore \quad S_1 + S_2 = S_1 + S_1(2^p - 1) = S_1[1 + (2^p - 1)] = S_1 2^p.$$

To sum $S_1$ we use the high school formula for the sum of a finite number of terms of a geometric progression,

$$S = \frac{lr - a}{r - 1}$$

$$S_1 = \frac{2^{p-1} \cdot 2 - 1}{2 - 1} = 2^p - 1.$$

Thus our total is

$$S_1 + S_2 = (2^p - 1)2^p = 2^p(2^p - 1)$$
$$= 2[2^{p-1}(2^p - 1)] = 2N.$$

This is the sum of the factors. Why did it come out $2N$ instead of $N$? Because we neglected to delete $N$ itself as one of the factors when we formed $S_2$. Thus the sum of the *proper* factors, including 1 but not including $N$, is $N$, and $N$ is therefore a perfect number.

The fact that $2^p - 1$ was prime had to be used in the proof; so the next question is, what numbers of the form $2^p - 1$ are prime? These are called Mersenne numbers after the seventeenth-century number theoretician. It is possible to prove that for $2^p - 1$ to be prime, it is *necessary* that $p$ be a prime, but *not sufficient*. We skipped the prime $p = 11$ for a very good reason: the Mersenne number $2^{11} = 1$ is composite (it equals $23 \times 89$), and therefore the number $2^{10}(2^{11}-1)$ is not perfect.

It is also possible to prove (Euler was the first to do it,) that not only some, but all even perfect numbers have this form. That is, an even number is perfect *if and only if* it is of the form

$$N = 2^{p-1}(2^p - 1), \text{ where } 2^p - 1 \text{ is prime.}$$

To search, then, for more even perfect numbers is to ask what Mersenne numbers are prime. To this difficult question there is no systematic answer. For a long time only 8 were known, and it is only since 1952 that 11 more (all huge) have been discovered, the latest in 1964.

O

A number the sum of whose proper divisors is less than the number itself is called *deficient,* and a number exceeded by this sum is called *abundant.* In 1962 the following problem was published.

"It is known that every number greater than 83160 can be expressed as the sum of two abundant numbers. Some lesser numbers cannot be so expressed. Inasmuch as 12 is the smallest abundant number, certainly no number less than 24 can be so expressed. What is the largest integer not so expressible?"

The problem had been completely solved for even numbers: 26, 28, 34, and 46 are the only even numbers greater than 24 that are not the sum of two abundant numbers. But for the odds the answer was unknown at that time. Although this is certainly not a very exciting problem, it is mentioned here because of the way in which it has been solved. T. R. Parkin and L. J. Lander hit upon it "as a test to try the capabilities of a very small digital computer (CDC 160-A) to do certain kinds of research in number theory." First by means of one page of analysis they reduced the lower limit to 28123. That is, it was possible to show that all numbers greater than 28123 can be expressed as the sum of two abundant numbers. There appeared to be no way of pushing this limit any lower by theoretical considerations.

They then programmed the computer to tabulate the decomposition of each odd number from 941 to 28999 into the sum of two abundant numbers if such a decomposition exists and to print out $0 + 0$ if it could not be so decomposed. There resulted 962 non-decomposables, of which the greatest was 20161, the sought solution. These results are described in a few typed pages. Then follows Appendix A, 91 pages of closely packed tabular material reproduced from the output sheets of the computer. From these pages it is easy to select the required data.

This was a small computer. Parkin and Lander say, "Indeed, this size problem is well within the range of hand calculation, but the use of a computer is quicker and more fun!" But similar methods can be applied to bigger and more serious problems with bigger computers. Computers cannot devise proofs or methods of solution. But they can aid in providing data in a form that may help immensely toward the solution of numerical problems, and they can turn out these data for hours on end with complete accuracy as fast as the printout mechanism will operate. Calculations whose mere size and tedium repel the mathematician are handled easily and rapidly by the computer. More data have thus been obtained concerning certain problems of very large numbers during the past decade than in all the previous years of the history of these problems.

# 3

# Prime numbers as building blocks

The Fundamental Theorem of Arithmetic states that a number can be factored into its prime factors in essentially only one way. The "essentially" means that $2 \times 5 \times 2$ is not to be considered a different factorization of 20 from $2 \times 2 \times 5$. No one can come along with some *other* primes, perhaps 7 and 3, that could also be multiplied together to produce 20. This is fairly obvious for small numbers; the theorem says that it holds for all numbers, large and small, and it is so important that it is worth proving.

We need two preliminary theorems, or "lemmas," before we can proceed. Someone once defined lemma as "the hard part of the proof," and he was not far off the mark. Often many pages of tough preliminary reasoning are necessary before a proof can be pushed to its conclusion.

If two numbers have only the factor 1 in common, they are said to be *relatively prime*. Note that two numbers do not have to be prime in order to be relatively prime. The two composite numbers $14 = 2 \times 7$ and $45 = 3 \times 3 \times 5$ are relatively prime, sharing no common factor (except 1).

*Lemma 1.* If several numbers are all relatively prime to a number $A$, then their product is relatively prime to $A$. This follows from the definition; for no new factors were introduced in the product, and there were none in common with $A$ to begin with.

*Lemma 2.* If a product of several numbers is divisible by a prime $p$, then at least one of them is divisible by $p$. For to say that none is divisible by $p$ is to say that each is relatively prime to $p$, and then the product would be relatively prime to $p$ by Lemma 1.

We can now prove the unique factorization theorem: every composite number $N$ can be expressed as a product of primes in (essentially) only one way. Suppose there are two factorizations:

$$N = p_1 \cdot p_2 \cdots p_m = q_1 \cdot q_2 \cdots q_n.$$

Note that at the outset we cannot assume that $m = n$. Now $q_1$ divides some one of $p_1 \cdots p_m$, by Lemma 2. But all the $p_i$ are primes; hence the only way $q_1$ can divide a $p$ is that $q_1 =$ that $p$, say $p_1$. (Relabel the $p_i$ if necessary: *one* of them is equal to $q_1$.) Now divide $p_1 = q_1$ out of both sides. We are left with

$$p_2 \cdot p_3 \cdots p_m = q_2 \cdot q_3 \cdots q_n.$$

Repeat the process until all the $q_i$ are gone. Then there can be no $p_i$ remaining either; for the product of *all* the original $q$'s was assumed to be $= N$, and therefore the whole value of $N$ has been divided out of the $p$ product. Thus we have identified each $q$ exactly with a $p$, which is what we set out to do.

Inasmuch as every number is either prime itself or can be broken down into its unique prime factors, but no further, the primes would seem to be endowed with a certain fundamental importance. If they are indeed the building blocks of the system they should appear in many formulas; one would expect all numerical calculations to be readily expressible in terms of primes. Unfortunately, this is far from true. Although many secrets seem to be locked up in the theory of prime numbers, so far we have been unable to find the keys. And some of the locks are so substantial that number theorists are beginning to think that perhaps there are no keys; that we have been asking the wrong questions; and that some of the properties of primes hitherto thought significant may be only trivial and co-incidental. We shall have occasion to return to this theme.

O

It is often of interest to find out whether two given numbers have any factors in common, and if so which ones. One way to find out would be to factor each number into its prime components and compare the two lists. But there is a better way, which comes down to us from Euclid with his name attached. This "Euclidean algorithm" tells us not only some common factor but all of them rolled into one in the form of the greatest common factor (or divisor) of the two numbers.

The operation of the Euclidean algorithm depends on the fact that if two numbers have a common factor $p$ (that is, are divisible by $p$,) then $p$ is also a divisor of their difference. The two numbers that have $p$ as a factor can be written $ap$ and $bp;$ but then their difference, $ap - bp = (a - b)p$, has the factor $p$ also.

Let us find the greatest common divisor of 108 and 30 by means of the algorithm. The difference between the two numbers is 78. Therefore any divisor of 78 also divides the other two, according to the previous paragraph; and we move to a consideration of 78 and 30 and start over again. The difference now is 48. One more difference, 48 − 30, is 18. Now we move to 30 − 18 = 12, then to 18 − 12 = 6, then 12 − 6 = 6, and finally 6 − 6 = 0. The last number before we arrive at the zero is the greatest common divisor: in this case, 6.

Two remarks should be made. First, there is a short cut. We had to subtract 30 from 108 three times in order to arrive at a number smaller than 30, namely, 18. But division is a short cut for repeated subtraction, just as multiplication is repeated addition. So why not *divide* 30 into 108, and take the *remainder*, 18, for our next divisor? This is what Euclid did. The work looks like this:

$$30 \overline{)108}$$
$$\qquad 3, \text{ remainder } 18$$
$$18 \overline{)30}$$
$$\qquad 1, \text{ remainder } 12$$
$$12 \overline{)18}$$
$$\qquad 1, \text{ remainder } 6$$
$$6 \overline{)12}$$
$$\qquad 2, \text{ remainder } 0, \text{ STOP.}$$

The second remark is more important: how do we know that 6 is the *greatest* common divisor just because it is the first divisor that yields a remainder of zero? Because the existence of any non-zero remainder means that the whole divisor is itself not a factor of the dividend. We know that our common factors are factors of both; the greatest factor of any divisor is certainly the divisor itself; and we stop as soon as we come to one that evenly divides the dividend. Had there been a larger one, it would have appeared sooner.

We mentioned earlier that the composite numbers 14 and 45 are relatively prime. This is disclosed by the Euclidean algorithm:

```
14 ) 45
      3, remainder 3
 3 ) 14
      4, remainder 2
 2 )  3
      1, remainder 1
 1 )  2
      2, remainder 0, STOP.  G.C.D. 1.
```

Surely it is a rather remarkable process that can tell us whether two numbers have any common factors *without factoring either of them.*

O

What is the probability that two numbers selected at random are relatively prime? If you could somehow put "all" the positive integers into a big hat, shuffle them, and pull out two, do you think that the chances would favor their being prime to each other or not? The question seems at first glance to be impossibly difficult to answer. That is why we are asking it.

The first hurdle is the "all." What do we mean by any two random numbers? The concept is not operational, because you cannot in fact put all the numbers into a hat. There are two ways to get around this. One is to say that we can imagine two numbers falling at random out of the whole aggregate of positive integers. If this does not please you (and there are many mathematicians it does not please in the least), we can actually do the problem for the first

few numbers (say 1 to 100, inclusive), and then do it over
for the numbers 1 to 1000, and so on. In a very few steps of
this kind we find that our answer is settling down, or as the
mathematician says, tending to a limit. And in this instance
the convergence toward the limit is rapid.

Before attacking our problem we must first assemble the
necessary ammunition. Among our weapons are some of
the ideas and principles of probability theory.

The number of ways to obtain a successful outcome of an
event, divided by the total number of outcomes of all kinds,
is called the probability of success for that event. There are
two outcomes in the toss of a coin—heads or tails. Only one
of these is a success if you are betting on heads. Thus the
probability of tossing a head on any one toss is said to be 1/2.
On the other hand, there are six faces on an ordinary cubical
die, so if we are trying to throw, say, a 1, the probability of
success is 1/6. (Note that we are not going to admit into
our experiments biased coins or loaded dice.)

Since an experiment must either succeed or not succeed,
the number of ways to succeed plus the number of ways to
fail equal the total number of outcomes of the experiment.
Thus if $p$ is the probability of success and $q$ is the probability
of failure, we have

$$p + q = \frac{\text{successes}}{\text{total number}} + \frac{\text{failures}}{\text{total number}}$$

$$= \frac{\text{successes} + \text{failures}}{\text{total number}} = 1.$$

That is,

$$p + q = 1, \quad \text{or} \quad q = 1 - p.$$

Next we have to know something about "the joint probability of independent events." If the probability of tossing

a head with one coin is 1/2, what is the probability of tossing two heads with two coins? If we shake two coins in a cup and then toss them out on the table, the way each falls is assumed to be independent of the other. So we simply look again at the total number of ways of getting results. There are three possible outcomes:

$$T\ T, \quad \text{or} \quad H\ T, \quad \text{or} \quad H\ H.$$

Since only one of these is a success (two heads), the probability of tossing two heads must be 1/3? Wrong. You have already detected the error, of course: we forgot to list a fourth possibility, *TH*, which is *not* the same as *HT* (label the coins if you have to). Therefore the correct probability is not 1/3 but 1/4.

Another way of doing the experiment is to toss one coin, record the result, and then pick it up and toss it again. The probability of obtaining a head on the first toss is 1/2; but then, only half the time *after that* will you get another head, so the product is wanted: $\frac{1}{2} \times \frac{1}{2} = \frac{1}{4}$.

What if we tossed three coins (or one coin three times)? The possible outcomes would be

$$T\ T\ T \quad T\ T\ H \quad H\ H\ T \quad H\ H\ H$$
$$T\ H\ T \quad H\ T\ H$$
$$H\ T\ T \quad T\ H\ H$$

Since only one of eight is three heads, the probability of tossing heads three times is 1/8.

Considerations like these lead to the formation of the law stating that the joint probability of $n$ independent events is the product of their separate probabilities. We could have arrived at the 1/8 by multiplying together the probability of getting each head separately: $\frac{1}{2} \times \frac{1}{2} \times \frac{1}{2} = 1/8$.

The biggest gun in our arsenal is this rather astonishing equation:

$$\left(1 + \frac{1}{2^2} + \frac{1}{3^2} + \frac{1}{4^2} + \frac{1}{5^2} + \cdots\right)\left(1 - \frac{1}{2^2}\right)\left(1 - \frac{1}{3^2}\right)$$

$$\times \left(1 - \frac{1}{5^2}\right)\left(1 - \frac{1}{7^2}\right)\left(1 - \frac{1}{11^2}\right) \cdots = 1$$

The first parenthesis contains the infinite series whose terms are the reciprocals of the squares of all the natural numbers. That this series converges can be proved by use of the "integral test." In fact it is known to converge to the value $\frac{\pi^2}{6}$.

It is a pity that we do not have at our disposal the mathematical tools necessary to prove this interesting claim. It can be done with Fourier series, a topic well beyond the range of this book. We ask you to take our word for it (or look up the reference given in the Notes) because we need the result to solve our problem.

Looking again at the equation we see that each parenthesis after the long one contains a reciprocal of a *prime* square. To show that this infinite product does indeed yield the value 1, we start multiplying:

$$1 + \frac{1}{2^2} + \frac{1}{3^2} + \frac{1}{4^2} + \frac{1}{5^2} + \frac{1}{6^2} + \cdots$$

$$1 - \frac{1}{2^2}$$

$$\overline{\rule{0pt}{0pt}\hspace{8cm}}$$

$$1 + \frac{1}{2^2} + \frac{1}{3^2} + \frac{1}{4^2} + \frac{1}{5^2} + \frac{1}{6^2} + \cdots$$

$$\quad\ - \frac{1}{2^2} \qquad\quad - \frac{1}{4^2} \qquad\quad - \frac{1}{6^2} - \cdots$$

$$\overline{\rule{0pt}{0pt}\hspace{8cm}}$$

$$1 \qquad\quad + \frac{1}{3^2} \qquad\quad + \frac{1}{5^2} \qquad\quad + \cdots$$

It is evident that all the multiples of $\frac{1}{2^2}$ have been knocked out by the first factor.  Continue:

$$1 + \frac{1}{3^2} + \frac{1}{5^2} + \frac{1}{7^2} + \frac{1}{9^2} + \cdots$$

$$1 - \frac{1}{3^2}$$

$$\overline{1 + \frac{1}{3^2} + \frac{1}{5^2} + \frac{1}{7^2} + \frac{1}{9^2} + \cdots}$$

$$-\frac{1}{3^2} \qquad\qquad -\frac{1}{9^2} - \cdots$$

$$\overline{1 \qquad + \frac{1}{5^2} + \frac{1}{7^2} \qquad + \frac{1}{11^2} + \cdots}$$

There go all the multiples of $\frac{1}{3^2}$.  We do not need the multiples of $\frac{1}{4^2}$ because they have already been caught as multiples of $\frac{1}{2^2}$.  We need only to take care of the next *primes*:  multiples of $\frac{1}{5^2}, \frac{1}{7^2}$, and so on.  After a few steps we are easily convinced that we are left, each time, with 1 plus "a very small quantity"; that the very small quantity becomes smaller at each step; and that we can make it as small as we please by taking a sufficient number of steps.  The mathematician summarizes all this by saying that the "partial products" approach 1 as a limit; and this in turn is all he means when he states that the infinite product *is* 1.

We can say, then, that

$$\left(1 - \frac{1}{2^2}\right)\left(1 - \frac{1}{3^2}\right)\left(1 - \frac{1}{5^2}\right) \cdots = \frac{1}{\frac{\pi^2}{6}} = \frac{6}{\pi^2}.$$

At last we are ready to proceed with the solution of the original problem and with all our weapons lined up, the conquest will be quite easy.

Let $m$ and $n$ be the given random numbers and let $a$ be any prime. Now $1/a$ of all the integers are divisible by $a$, because $a$ divides every $a$'th one. Hence the probability that $a$ divides $m$ is $1/a$. Likewise the probability that $a$ divides $n$ is $1/a$. Hence the probability that $a$ divides *both* $m$ and $n$ is the joint probability $\dfrac{1}{a} \times \dfrac{1}{a} = \dfrac{1}{a^2}$. Therefore the probability that $a$ does *not* divide both $m$ and $n$ is $1 - \dfrac{1}{a^2}$. (You should recognize $q$ of our preliminary discussion.)

Now the probability $P$ that $m$ and $n$ are relatively prime means this probability $\left(1 - \dfrac{1}{a^2}\right)$ occurring for *all* primes $a$. Thus, applying the joint law again,

$$P = \left(1 - \frac{1}{2^2}\right)\left(1 - \frac{1}{3^2}\right)\left(1 - \frac{1}{5^2}\right) \cdots = \frac{6}{\pi^2}$$

= approximately .61, if anyone is interested in a decimal value.

If we were to select two numbers at random from a finite collection and adjust the probabilities accordingly we would find that, as the collection increased in size, the probability would rapidly approach .61.

We have discussed this problem at some length because it illustrates how a question quite easy to ask can in fact lead one far afield in quest of the answer. Our problem, asking about common divisors, turned out to require much more than just number theory for its solution. We had to delve into probabilities; we had to know something about convergence of infinite series and the multiplication of an infinite product; and we needed a value of the Riemann Zeta-function, the technical name for the series that converged to $\pi^2/6$.

○

In the last section we assumed something that we have not yet proved: that the number of primes is infinite. The string of parentheses in the expression for $P$ was not supposed to terminate. How do we know that? Why are we sure that we will never run out of primes?

One way to show that a set contains an infinite number of elements is to exhibit a rule or law of formation whereby, from any element, one can obtain a larger one (usually the *next* larger one). Thus the set of positive integers is infinite because, given any number $N$ however large, there is always another larger, namely, $N + 1$.

Such a *constructive* proof is not available to us, at the present state of knowledge, for primes. It would be pleasant to be able to turn to a formula, or even a recursion relation, that would grind out only primes. Unfortunately, we are obliged to do without this luxury. Nevertheless we can prove that the number of primes is infinite, by the device of showing that there is no largest one. For a finite collection has always *some* largest member.

The proof proceeds by contradiction. Assume for the moment that there is a largest prime, say $N$. Then form the number $Q = N! + 1$. The symbol $N!$ means

$$N(N - 1)(N - 2)(N - 3) \cdots 3 \times 2 \times 1$$

For instance, $5! = 5 \times 4 \times 3 \times 2 \times 1 = 120$. Now look at $5! + 1$. It certainly cannot be evenly divisible by any of the numbers 2, 3, 4, or 5: there will be a remainder of 1 in every case, because 5! *is* evenly divisible by each. Therefore one of two things: either 5! + 1 is *prime* (it happens not to be), or 5! + 1 is divisible by some prime greater than 5. (It is divisible by 11.) By exactly this reasoning, $Q$ is either a prime (certainly greater than $N$) or else divisible by a prime greater than $N$. In either case, $N$ is not the greatest prime, and our assumption that there exists a largest prime has been proven false.

We said just now that there is no known formula that produces all the primes. Mathematicians would be happy if they could find something much more modest: a procedure that would be guaranteed to produce primes *at all* (which is very different from *all the primes*). Fermat thought that he might have such a formula:

$$F_n = 2^{2^n} + 1.$$

Substituting successively 0, 1, 2, 3, and 4 for $n$ yields:

$$F_0 = 2^{2^0} + 1 = 3$$
$$F_1 = 2^{2^1} + 1 = 5$$
$$F_2 = 2^{2^2} + 1 = 17$$
$$F_3 = 2^{2^3} + 1 = 257$$
$$F_4 = 2^{2^4} + 1 = 65537.$$

These are indeed all primes; but now the process breaks down, for

$$F_5 = 2^{2^5} + 1 = 4294967297$$

is composite. It is hardly surprising that Fermat happened not to find the factors: 641 and 6700417. The higher Fermat numbers have been the subject of prolonged study, and to date no more primes have been found among them.

A formula that produces *a few* primes is

$$Y = x^2 - x + 41.$$

This turns out a $Y$ that is a prime if you put $x = 1$, or 2, or 3, all the way up to $x = 40$. But $x = 41$ yields the composite number $Y = 41^2$. A similar formula, to which we alluded on page 14, is

$$Y = x^2 - 79x + 1601.$$

As remarked, it fails for the first time for $x = 80$, and many times thereafter.

Indeed, *no* polynomial in $x$ can ever produce primes for all $x$. To prove this statement we assume the contrary and observe what happens.

Suppose such a polynomial exists:

$$Y(x) = a_0 + a_1x + a_2x^2 + \cdots + a_nx^n.$$

The $a$'s are any integers, positive, negative, or zero, and their subscripts are merely labels. The superscripts on the $x$'s are, as usual, powers. $Y(x)$ means the value of the polynomial when $x$ is substituted into it. Since it is assumed to produce primes for all $x$, select some $x$, say $x = b$, that yields a prime $p$: $Y(b) = p$. We now show that $p$ divides $Y(b + mp)$ for any $m = 1, 2, 3, \ldots$

$$Y(b + mp) = a_0 + a_1(b + mp) +$$
$$a_2(b + mp)^2 + \cdots + a_n(b + mp)^n.$$

Expanding the parentheses and separating out the first term from each, we get

$$Y(b + mp) = [a_0 + a_1b + a_2b^2 + \cdots + a_nb^n] +$$

[many other terms, all of which contain $p$ as a factor].

But the first bracket is exactly $Y(b) = p$. Factoring out $p$ from the second bracket, we can write

$$Y(b + mp) = p + p \times \text{(another polynomial in } b, m, p).$$

As $m$ takes on infinitely many integer values, so must this new polynomial in $m$, of degree $n$; for no polynomial (except a constant) can assume the same value more than $n$ times. Thus we have an infinite number of values of $x$, namely each $(b + mp)$, such that when put through the original polynomial they produce values of $Y$ divisible by $p$, contradicting the hypothesis. Hence no such prime-producing polynomial exists.

# 4
## Congruence arithmetic

Congruence arithmetic is in common use in the building industry. Designers, manufacturers, even carpenters and masons, without ever having studied number theory, use it every day. Modern industrial methods have introduced the once abstract and purely theoretical notion of a *modulus* for entirely practical economic reasons.

If a prefabricated wall comes in 8-foot sections the builder speaks of this as a *module* and seeks, for the various elements of the construction, lengths that will fit the 8-foot module. The architect plans the lengths of all his walls, wherever possible, to come out in multiples of 8. Tiles, windows, and all the rest can be assembled with the least amount of trouble if they fit the module.

It may not be possible, however, to build the whole house this way. Suppose the carpenter looks at the plans and sees a length of 11-foot wall, and at another place he finds a 19-foot wall. He notes that one 8-foot section will leave 3 feet over for the 11 foot wall, and two 8-foot sections can be used for the 19 foot span, but likewise will leave 3 feet unaccounted for. Therefore the problem to be dealt with in both cases is the same. It makes no difference how many modular sections are to be used; that's the easy part. What must be filled in with some kind of hand work is the space of three feet—the *remainder* on division by 8. The fact that the carpenter has the identical problem in both cases is expressed in number-theoretic language by saying that 11 and 19 are *congruent* modulo 8:

$$19 \equiv 11 \ (\text{mod } 8),$$

read "nineteen is congruent to eleven modulo eight." Note the key role played by the remainder. Once the modulus is established as a divisor, it is *equality of the two remainders* that makes two numbers congruent.

To make sure that you feel at home with congruences, which we intend to use freely, we offer a second illustration.

If a scientist is performing an experiment in which it is necessary for him to keep track of the total number of hours that have elapsed since the start of the experiment, he may label the hours sequentially 1, 2, 3, etc. When 41 hours have elapsed, it is 41 o'clock "experiment time." How does he reduce experiment time (e.t.) to ordinary time? If zero hours e.t. corresponded to midnight, his task is easy: he simply divides by 12 and the *remainder* is the time of day. 41 e.t. is thus 5 o'clock, because 12 goes into 41 with a remainder of 5; 41 is congruent to 5 modulo 12:

$$41 \equiv 5 \ (\text{mod } 12).$$

For the purpose of telling the time of day it is not necessary to know *how many* times 12 is contained in 41, but only the remainder, 5. Of course if one wants to distinguish between a.m. and p.m., then it would be better to divide by 24. We then find that 41 is congruent to 17 modulo 24 (24 goes once into 41 with a remainder of 17). This means that 41 e.t. is 17 hours after midnight. But $17 \equiv 5 \pmod{12}$, which means that 17 after midnight is 5 p.m. and unless we are in the navy or reading a French railway timetable we do not customarily say 17 o'clock. In fact we simply reduce all our afternoon hours modulo 12.

Another thing that $41 \equiv 5 \pmod{12}$ means is, of course, that $(41 - 5)$ is a multiple of 12: take away the remainder and it would come out even. But if so, then $(41 - 5)k$ is also a multiple of 12 for any $k$. But this says that $41k - 5k$ is a multiple of 12, or in other words,

$$41k \equiv 5k \pmod{12}.$$

We begin to see that it is with good reason that the congruence sign has been chosen to resemble the equals sign. One can multiply both sides of a congruence by the same number just as one can multiply both sides of an equation by the same number:
If

$$a \equiv b \pmod{m}$$

then

$$ak \equiv bk \pmod{m}.$$

In ordinary equalities we can multiply equals by equals and add equals to equals. It would be nice to know whether congruences follow these same rules.
Given

$$a \equiv b \pmod{m}$$

and

$$c \equiv d \pmod{m}$$

is it true that

$$ac \equiv bd \pmod{m}?$$

The question is, is $ac - bd$ divisible by $m$?

Now

$$c \equiv d \ (\text{mod} \ m)$$

says that $c - d = km$ for some $k$, or $c = km + d$. Therefore,

$$
\begin{aligned}
ac - bd &= a(km + d) - bd \\
&= akm + ad - bd \\
&= akm + d(a - b).
\end{aligned}
$$

But $(a - b)$ is divisible by $m$; that's what $a \equiv b \ (\text{mod} \ m)$ asserts. Therefore $ac - bd$ has $m$ as a factor, or $ac \equiv bd$ $(\text{mod} \ m)$.

The addition of equals to equals goes even more easily: Given

$$a - b = jm \text{ for some } j$$

and

$$c - d = km \text{ for some } k$$

$$(a + c) - (b + d) = (j + k)m = lm \text{ for some } l, \text{Q.E.D.}$$

We shall make use of one more property, whose proof is slightly more elegant than the two just given.
If

$$a \equiv b \ (\text{mod} \ m)$$

then

$$a^k \equiv b^k \ (\text{mod} \ m).$$

That is, we can raise both sides of a congruence to the same integral power.

*Proof.* $a^k - b^k$ is always divisible by $a - b$. But $a - b$ is divisible by $m$ (given). Therefore so is $a^k - b^k$. This completes the proof.

If you don't believe that $a^k - b^k$ is always evenly divisible by $a - b$, try it by long division. The quotient is the so-called "cyclotomic" expression,

$$a^{k-1} + a^{k-2}b + a^{k-3}b^2 + \cdots + a^2b^{k-3} + ab^{k-2} + b^{k-1}$$

of degree $k - 1$.

The relation of congruence is, like ordinary equality, an *equivalence relation*. This means that the relation must have three properties:

(1) Reflexivity:   $a \equiv a \pmod{m}$

(2) Symmetry:  if $a \equiv b \pmod{m}$, then $b \equiv a \pmod{m}$

(3) Transitivity: if $a \equiv b \pmod{m}$ and $b \equiv c \pmod{m}$,
then $a \equiv c \pmod{m}$.

When a number is divided by $m$, there are only $m$ possible remainders, if we include 0. If we identify every integer, positive, negative, or zero, with its remainder modulo $m$, we thus have all the integers *partitioned* into *congruence classes*, or *residue classes*, modulo $m$. It is this reduction of the infinite class of integers to a comparable finite class that gives modulo arithmetic its very considerable power. For instance, since 8, 15, 22, 29, etc., and also $-6$, $-13$, etc. are all congruent to 1 (mod 7), they are members of the same residue class modulo 7. Another way of saying this is that when we count beyond 7 we just start over again. In arithmetic modulo 7, 8 is equivalent to 1 in many senses.

We have an everyday example of all this in clock arithmetic, where counting starts over again every 12 hours. We don't have much occasion to add or multiply times of day. If we did, we should need special addition and multiplication tables; but instead of extending indefinitely, they would each consist only of a $12 \times 12$ square. For brevity, we exhibit an addition table and a multiplication table for modulo 5 arithmetic.

| Addition mod 5 | | | | | |
|---|---|---|---|---|---|
| | 0 | 1 | 2 | 3 | 4 |
| 0 | 0 | 1 | 2 | 3 | 4 |
| 1 | 1 | 2 | 3 | 4 | 0 |
| 2 | 2 | 3 | 4 | 0 | 1 |
| 3 | 3 | 4 | 0 | 1 | 2 |
| 4 | 4 | 0 | 1 | 2 | 3 |

| Multiplication mod 5 | | | | | |
|---|---|---|---|---|---|
| | 0 | 1 | 2 | 3 | 4 |
| 0 | 0 | 0 | 0 | 0 | 0 |
| 1 | 0 | 1 | 2 | 3 | 4 |
| 2 | 0 | 2 | 4 | 1 | 3 |
| 3 | 0 | 3 | 1 | 4 | 2 |
| 4 | 0 | 4 | 3 | 2 | 1 |

In number theory one must often work with very large numbers. If these can be reduced to equivalent smaller numbers, much time-consuming labor can be avoided. Herein lies one of the great contributions of modulo arithmetic.

EXAMPLE 1. Is 999999 evenly divisible by 7? We will find later that this is not such an empty question as it may appear. Of course we could do the short division without much trouble; but let us instead use the congruence idea:

$$999999 = 10^6 - 1$$
$$10 \equiv 3 \pmod 7$$
$$\therefore \quad 10^6 \equiv 3^6 \pmod 7$$

But $3^6 = (3^2)^3 = 9^3$, and $9 \equiv 2 \pmod 7$
$$\therefore \quad 9^3 \equiv 2^3 \equiv 1 \pmod 7.$$

Finally, therefore, substituting from above,

$$10^6 \equiv 1 \pmod 7.$$

That is to say, 999999 *is* evenly divisible by 7.

EXAMPLE 2. "Every even power of any odd number is congruent to 1 modulo 8." This modulus partitions the integers into 8 residue classes, all the odd numbers being congruent to one of 1, 3, 5, 7, by definition. If we square both sides of these four possible congruences, we have that every odd square is congruent to one of the numbers 1, 9, 25, or 49 (mod 8). But these all happen to be $\equiv 1$ (mod 8). Raising both sides to a power (to obtain any even power) will not change this congruence.

EXAMPLE 3. We claimed in Chapter 3 that the fifth Fermat number,

$$F_5 = 2^{2^5} + 1 = 2^{32} + 1,$$

is divisible by 641. Let us prove this without any gruesome arithmetic. We are not eager to raise 2 to the 32nd power.

$$640 = 10 \times 64 = 5 \times 128 = 5 \times 2^7 \equiv -1 \text{ (mod 641)}.$$

Taking the fourth power of both sides,

$$5^4 \times 2^{28} \equiv 1 \text{ (mod 641)}.$$

But $5^4 = 625 \equiv -16$ (mod 641), which we can write $5^4 \equiv -(2^4)$ (mod 641). But two numbers having the same remainder belong to the same equivalence class, and one may be replaced by the other without destroying the congruence (see Notes). This says that we can write

$$-(2^4) \times 2^{28} \equiv 1 \text{ (mod 641)}$$
$$-(2^{32}) \equiv 1 \text{ (mod 641)}$$
$$2^{32} \equiv -1 \text{ (mod 641)}$$
$$2^{32} + 1 \equiv 0 \text{ (mod 641)}.$$

○

When we were demonstrating the strong parallel between congruences and equations there was one property of equalities that we were careful not to mention: "equals may be divided by equals." We ask now, under what

conditions, if any, may a constant be divided out of both sides of a congruence?
Given

$$ab \equiv ac \pmod{m},$$

when is

$$b \equiv c \pmod{m}?$$

Reverting, as usual, to the meaning of the congruence, we are given that

$(ab - ac)$ is divisible by $m$; or,
$a(b - c)$ is divisible by $m$.

What we want to conclude is that

$(b - c)$ is divisible by $m$.

This is certainly true if $m$ and $a$ are relatively prime; for then there is no other way for $m$ to divide the product $a(b - c)$. If $a$ is not relatively prime to $m$, we do not know: $m$ might or might not divide $(b - c)$.

EXAMPLES. (a)  $99 \equiv 9 \pmod{10}$
and  $11 \equiv 1 \pmod{10}$.

We divided by 9, a number relatively prime to the modulus.

(b)  $48 \equiv 12 \pmod{6}$
and (1)  $8 \equiv 2 \pmod{6}$
but (2)  $4 \not\equiv 1 \pmod{6}$

(The sign $\not\equiv$ means "is not congruent to," just as $\neq$ means unequal.) In both (1) and (2) of case (b) we divided by a number not prime to the modulus. Once it worked and once it didn't. It is not hard to develop a further rule for case (b), and we leave this to you. We are going to be more concerned with case (a), where the division is surely allowed.

We can now prove a theorem of Fermat's: If $a$ is not divisible by a prime $p$, then $a^{p-1} \equiv 1 \pmod{p}$.

*Proof.* We consider the set of numbers $a$, $2a$, $3a$, ... $(p - 1)a$, and ask how they fall into the residue classes modulo $p$. Remember that $p$ partitions all the integers into exactly $p$ residue classes. But no one of our set is congruent to $p$ itself (because we stopped at $(p - 1)a$, and $a$ is not divisible by $p$). Furthermore, no two of the set are congruent to each other (that is to say, no two of them belong in the same residue class). For if $xa \equiv ya \pmod{p}$ then $x \equiv y$ $\pmod{p}$ by our division theorem. Therefore each is congruent to some one of 1, 2, 3, ... $(p - 1)$ (not necessarily in order). The situation is the same as that displayed in the table of multiplications modulo 5. Therefore the product of all the set is congruent to the product $(p - 1)!$. That is, factoring out the $a$'s,

$$a^{p-1}(p - 1)! \equiv (p - 1)! \pmod{p}.$$

And now we know that $(p - 1)!$ is not divisible by $p$, so that we can divide it out to get

$$a^{p-1} \equiv 1 \pmod{p}.$$

If we had had this theorem three pages back we need not have troubled to prove that

$$10^6 \equiv 1 \pmod{7}.$$

Fermat's Theorem says that it is true.

○

We now take a look at the binomial expansion,

$$(x + y)^0 = 1$$
$$(x + y)^1 = 1x + 1y$$
$$(x + y)^2 = 1x^2 + 2xy + 1y^2$$
$$(x + y)^3 = 1x^3 + 3x^2y + 3xy^2 + 1y^3$$
$$(x + y)^4 = 1x^4 + 4x^3y + 6x^2y^2 + 4xy^3 + 1y^4,$$

and so on. It is easy to see how the exponents of $x$ and $y$ are shaping up. The question is how to predict the multipliers. These are the famous *binomial coefficients*, whose law of formation is well known. It is not hard to obtain the next row of coefficients from any given row. Writing only the coefficient part of each term in a skeletal multiplication,

| | | | | | | |
|---|---|---|---|---|---|---|
| $(x+y)^4$ | 1 | 4 | 6 | 4 | 1 | |
| $(x+y)$ | 1 | 1 | | | | |
| | 1 | 4 | 6 | 4 | 1 | |
| | | 1 | 4 | 6 | 4 | 1 |
| $(x+y)^5$ | 1 | 5 | 10 | 10 | 5 | 1 |

All this can be schematically arranged in Pascal's Triangle:

```
                              1
                          1       1
                      1       2       1
                  1       3       3       1
              1       4       6       4       1
          1       5      10      10       5       1
      1       6      15      20      15       6       1
  1       7      21      35      35      21       7       1
1     8      28      56      70      56      28       8       1
1   9    36    84   126   126    84    36     9     1
.   .    .     .     .     .     .     .     .     .     .
```

FIGURE 2. PASCAL'S TRIANGLE

Every number in each row of the triangle is obtained by adding the two numbers nearest it (diagonally) in the row above, except the two end numbers, which are always 1's.

The general expression for the $(k + 1)$th term in the expansion of the $n$'th row—that is, the $(k + 1)$th coefficient in $(x + y)^n$—is

$$\frac{n!}{k!\,(n - k)!}.$$

This represents the totality of combinations of $n$ things taken $k$ at a time, because that is the possible number of ways to obtain the combination of $x$'s and $y$'s with the right exponents to go with this coefficient. Because these things lie in the realm of elementary algebra rather than number theory we omit proofs, hoping that you may remember some of them from your high school days.

The binomial coefficients have many interesting properties. Suppose someone asks, is there a pair of adjacent coefficients (in any row, anywhere,) such that one is $\frac{2}{3}$ of its neighbor? Looking in the triangle we soon find that 4 and 6 in the fourth row meet this condition. Are there by any chance others? We do not see any at first glance except the matching pair 6, 4, in the same row. So we state the question in general form: when can we have the $(k + 1)$th term equal to $a$ times the $k$'th? This means

$$\frac{n!}{k!\,(n - k)!} = a\,\frac{n!}{(k - 1)!\,(n - k + 1)!},$$

which reduces to

$$n = ka + k - 1.$$

This is a surprise. It tells us that there are many answers to our question. Indeed, for *any* fraction $a = p/q$ there is an infinite number of consecutive pairs, one for each member of the residue class congruent to zero modulo $q$. For example, for $a = \frac{3}{2}$ we need only take $k \equiv 0 \pmod 2$. If $k = 2$,

$n = 4$, and this is the pair we have already picked up: 6, 4 in the fourth row. The next is $k = 4$, $n = 9$, and we find that $126 = \frac{3}{2} \times 84$ in the 9th row. The next pair is 3003, 2002 in the 14th row, and so on. That we can meet this demand in an infinity of different places in the triangle for any fraction $p/q$ whatever is somewhat unexpected.

A perhaps more significant property of the triangle is that all the numbers in the $n$'th row (except the first and the last, which are 1's) are evenly divisible by $n$ if and only if $n$ is a *prime*. Thus 5 and 7 divide each number in their respective rows but 8 and 9 do not. A glance at the general coefficient

$$C = \frac{n!}{k! \, (n - k)!}$$

is sufficient to show that if $n$ is a prime it must divide all but the first and last terms. For in the cancellation necessary to reduce the fraction to the integer $C$, nothing can cancel out the $n$ contained in $n!$, which thus survives as a factor of $C$. The converse is also true (see Notes for proof): if $n$ is composite it cannot divide *all* the coefficients $C$. This means it may divide *some* of them. How to tell which ones, or even how many, in each row, is a challenging unsolved problem.

Fermat's Theorem is closely connected with these questions, and we now give an alternate proof of the theorem, using induction on $a$.

First we observe that
if
$$a^{p-1} \equiv 1 \pmod{p}$$
then
$$a^{p} \equiv a \pmod{p}$$

That is, an equivalent statement of the theorem is, "If $p$ is a prime, then $a^{p} - a$ is divisible by $p$."

We start the induction by observing that the theorem is certainly true for $a = 1$: $1 \equiv 1 \pmod{p}$. Now make the induction assumption: suppose it is true that $a^p - a$ is divisible by $p$ for some $a$, say $a = b$; does it follow that it is true for $b + 1$? That is, does $p$ divide $(b + 1)^p - (b + 1)$? Expanding $(b + 1)^p$ by the binomial expansion, we have

$$(b + 1)^p - (b + 1)$$
$$= \{b^p + [\text{all terms except the first and the last}] + 1\}$$
$$- (b + 1)$$
$$= b^p - b + [\text{all terms except the first and the last}].$$

But now we know that, because $p$ is a prime, it divides each of the terms in the bracket. It also divides $b^p - b$ by the induction assumption. Therefore it divides the whole right-hand side, and hence must divide the left-hand side, which completes the proof.

○

Inasmuch as Fermat's Theorem can be stated, "If $p$ is a prime it divides $a^p - a$," one would naturally ask next about the converse: "If $a^p - a$ is divisible by $p$, then $p$ is a prime." If this were true it would be a criterion for primality. Strangely enough this converse holds *most* of the time *but not always*. Look, for instance, at the case $a = 2$. Now $2^n$ is the sum of all the numbers in the $n$'th row of the Pascal Triangle (easily provable). Hence $2^n - 2$ is the sum of all except the first and the last. We know that this sum *is* divisible by $n$ if $n$ is prime, because in that event each term of it is divisible by $n$. The question is, could it ever happen that the sum is divisible by $n$ even though $n$ is composite? The answer is yes, and it is apparently unpredictable when this will happen. If we examine the first few cases it seems as if no composite $n$ is going to divide $2^n - 2$; and in fact it is well known that the very first exception occurs at $n = 341 = 11 \times 31$. No wonder the converse of Fermat's Theorem looks like a sure thing if we judge only by appearance, for we are hardly likely to hit upon $2^{341} - 2$ in our wanderings: it is a number of more than a hundred digits!

We mention this principally because it affords a good example of the power of modulo arithmetic. It would be out of the question actually to calculate $2^{341} - 2$ and then divide it by 341 to find out whether the result came out even. But observe:

$$2^5 \equiv 1 \ (\text{mod } 31)$$
$$\therefore \quad (2^5)^{68} \equiv 1^{68} \ (\text{mod } 31);$$

or

$$2^{340} \equiv 1 \ (\text{mod } 31).$$

Also

$$2^5 \equiv -1 \ (\text{mod } 11),$$

hence

$$2^{340} \equiv 1 \ (\text{mod } 11).$$

Thus both 11 and 31 are prime factors of $2^{340} - 1$. And now

$$2^{341} - 2 = 2 \times (2^{340} - 1)$$
$$= 2 \times 11 \times 31 \times \text{other things}$$
$$= 2 \times 341 \times \text{other things}.$$

○

Some of our readers may already have come to grips with the famous "Monkey and Coconuts" problem, which has tantalized many a problem solver. We shall not only state it but will also solve it for you, although this seems almost like cheating.

Five sailors plan to divide a pile of coconuts among themselves in the morning. During the night one of them wakes up and decides to take his share. After throwing a coconut to a monkey to make the division come out even, he takes one fifth of the pile and goes back to sleep. The other four sailors do likewise, one after the other, each throwing a coconut to the monkey and taking one fifth of the remaining pile. In the morning the five sailors throw a coconut to the monkey and divide the remaining coconuts into five equal piles. What is the smallest number of coconuts that could have been in the original pile?

*Solution.* If the original number of coconuts is $N$ and the number each sailor receives in the final division is $a$, then

$$\tfrac{1}{5}(\tfrac{4}{5}(\tfrac{4}{5}(\tfrac{4}{5}(\tfrac{4}{5}(\tfrac{4}{5}(N-1)-1)-1)-1)-1)-1) = a.$$

Removing the parentheses and regrouping (with care!) yields

$$(\tfrac{4}{5})^5 N - [1 + \tfrac{4}{5} + (\tfrac{4}{5})^2 + (\tfrac{4}{5})^3 + (\tfrac{4}{5})^4 + (\tfrac{4}{5})^5] = 5a.$$

Summing the series in the bracket by means of the same formula for geometric progression that we used in deriving Euclid's form for perfect numbers, and clearing of fractions, we eventually arrive at

$$4^5(N+4) = 5^6(a+1).$$

Now regardless of the nature of $(a+1)$, the unique factorization theorem tells us that all the prime factors of $5^6 = 5 \times 5 \times 5 \times 5 \times 5 \times 5$ must also be contained in the left-hand side. Since none of them are in $4^5$, $5^6$ must divide $(N+4)$:

$(N+4)$ is divisible by $5^6 = 15625$.

That is,

$$N \equiv -4 \pmod{15625}.$$

But we can't start with $-4$ coconuts, and the very next integer in this residue class modulo 15625 is $-4 + 15625 = 15621$.

That's a lot of coconuts!

# 5

# Irrationals and iterations

The infinite sequence of the perfect squares begins like this:

1, 4, 9, 16, 25, 36, 49, 64, . . .

It is perhaps not immediately apparent that nowhere in this sequence is there one number that is twice another. No matter how far out we look, we can never find a perfect square that is double another perfect square. For if we could, the quotient would be 2. But the prime factors of squares appear always in pairs; and if we were to cancel $\frac{a^2}{b^2}$ until the result were $\frac{2}{1}$, it would mean that *each pair* of the factors of $b^2$ had been matched with a *pair* of the factors of $a^2$ in the cancellation. How, then, could there be a lone 2 left over?

The fact that $\frac{a^2}{b^2}$ can never equal 2 for any integral $a$ and $b$ is another way of saying that $\sqrt{2} = \frac{a}{b}$ is an impossible equation with $a$ and $b$ integers. We say that $\sqrt{2}$ is *irrational* (non-ratio-nal).

55

For the same reason, no member of the infinite sequence of squares is ever 3 times another, nor 5 times another, nor any prime times another. Nor, for that matter, is one square ever 6 times another, even though 6 is not a prime. So $\sqrt{6}$ is irrational also. But can one square be 4 times another? Certainly, because if $\dfrac{a^2}{b^2} = \dfrac{4}{1}$, it means that $4 = 2 \times 2$ was left over after the cancellation, and this does *not* prevent $a^2$ from existing as a square. For instance, $\frac{100}{25} = 4$. $\sqrt{4}$ is a *rational* number.

Likewise no cube can be a prime multiple of any other cube, and so on. Numbers like $\sqrt[3]{2}$, $\sqrt[3]{3}$, $\sqrt[3]{4}$, are all irrational.

The rationals are the numbers that have a fractional representation: they can be written as the quotient of two whole numbers. We now investigate some problems concerning the rationals.

○

A pattern whose meaning is not at first apparent is:

$$1 \times 142857 = 142857$$
$$2 \times 142857 = 285714$$
$$3 \times 142857 = 428571$$
$$4 \times 142857 = 571428$$
$$5 \times 142857 = 714285$$
$$6 \times 142857 = 857142$$
$$7 \times 142857 = 999999.$$

Why are the *cyclic permutations* of the same six numbers occurring as the integral multiples of 142857? Certainly not all numbers act this way. And why do the 9's suddenly put in an appearance?

The explanation is not far to seek. Suppose we decimalize $\frac{1}{7}$ by short division:

$$7)\overline{1.0^30^20^60^40^50\ ^10^30 \ldots}$$
$$.1\ 4\ 2\ 8\ 5\ 7, 1\ 4 \ldots$$

The first 1 after the decimal point in the "answer" means that 7 was contained once in 10. The *remainder* was 3, making the next division 7 into 30. This time the quotient was 4 and the remainder 2; and so on. The remainders are the crucial things: after at most six divisions we run out of new remainders, for there are only 6 different possible remainders under division by 7. And as soon as we arrive back at a remainder of 1, the whole process repeats. The decimal representation of $\frac{1}{7}$ has a repeating period six digits long.

If we decimalize $\frac{2}{7}$ we get the same period, only the division starts at a different place. The same can be said of $\frac{3}{7}, \ldots, \frac{6}{7}$. But $\frac{7}{7} = 1 = .999999,999999, \ldots$

Seven is a prime, and its reciprocal, $\frac{1}{7}$, has period $7 - 1 = 6$ digits long, the maximum possible length. But the period of $\frac{1}{3} = .33333 \ldots$ is only 1 digit long where it might have been 2, 3 being a prime. 13 is a prime, so that $\frac{1}{13}$ might have had period length 12, but in fact the division goes only 6 steps before the same remainder comes 'round again. Is 7, then, the only integer whose reciprocal has the maximum possible repetend? No indeed. The next one is 17:

$$\frac{1}{17} = .0588235294117647,0588 \ldots$$

The "full repetend" primes are not infrequent (there are seven more less than 100), but how to predict exactly which numbers behave this way is a question that has intrigued mathematicians for a long time. So far no one has come up with an answer. Even the great Gauss tackled the problem without solving it, although he was led through its study to some more important and far reaching results.

Fermat's Theorem of the last chapter tells us, that since 7 does not divide 10,

$$10^6 \equiv 1 \pmod{7}.$$

Omitting the decimal point from our short division gives us the division of $10^6$ by 7. Thus Fermat's Theorem could have told us just when that all-important remainder of 1, starting us around the circuit again, had to appear. $10^6 - 1$ = 999999 is evenly divisible by 7, so that instead of dividing 7 into 1 we could just as easily have divided it into a string of 9's until the remainder was zero. This is in fact an alternative way to find the repeating period of any prime. (For brevity we speak of the period of a number when in fact we mean the length of the period of the decimal expansion of its reciprocal.)

The difficulty with Fermat's Theorem is that it does not guarantee that 6 is the *smallest* exponent $e$ for which

$$10^e - 1 \equiv 0 \pmod{7}.$$

With 7 it happens to be. But, for instance,

$$10^{10} - 1 \equiv 0 \pmod{11},$$

and we find that

$$\tfrac{1}{11} = .090909 \ldots$$

The period, which cannot be *longer* than ten digits, is this time only 2 digits long. The theorem says that 11 must divide 9,999,999,999. It happens that 11 also divides the somewhat smaller number, 99.

One trifle we can predict. If $n$ is a prime, its period must end with that digit $d$ which is the "complement" of $n$ in the sense that the product $nd$ ends in 9. This is necessary in order for the subtraction at that stage to yield 1, so that the cycle can start over again. Examples: $d = 9$ for $\frac{1}{11}$, 7 for $\frac{1}{7}$, 3 for $\frac{1}{3}$, etc.

○

Eleven is a prime. 111 and 1111 are not. (a) Are there any other numbers consisting of a string of 1's that are prime, and if so (b) how many, and (c) how can we find them? The answer to (b) is unknown, and (c) is the kind of question for which we may quite possibly never have an answer.

The next two such primes are the numbers consisting of 19 and 23 1's, respectively. No others are known, nor even whether there are any more.

Suppose $p_1, p_2, \ldots, p_n$ are different primes all of which are known to have repeating periods $k$ digits long. We have just seen that this means that each of them divides the number $10^k - 1$, a string of $k$ 9's. But 9 also divides $10^k - 1$ for any $k$, leaving a string of $k$ 1's. Thus if $p_1, p_2, \ldots, p_n$ are known to be the *only* primes of period length $k$, then their product

$$p_1 \times p_2 \times \cdots \times p_n = \frac{10^k - 1}{9},$$

is the number consisting of exactly $k$ 1's. Thus the two problems are the same: if we knew all about the repeating decimals we would have at least partial answers to questions (b) and (c).

The situation is more fruitful in reverse. Suppose we ask, how many primes are there whose reciprocals have repeating decimals of period 7? This means that the prime itself need only divide 9,999,999, and it would appear to involve the formidable task of testing all primes up to $\sqrt{9,999,999}$. But note that all such primes will also have to divide 1,111, 111. If by any chance we happen to know that the *prime* factorization of 1,111,111 is 239 × 4649, we are in luck. This says that 239 and 4649 have periods of length 7, and that they are the *only such numbers*. Observing that we must guard against the trouble we had with $\frac{1}{11}$, whose period was 2, a *factor* of the maximum possible 10, we limit the present

| Period length | Primes |
|---|---|
| \multicolumn{2}{c}{List of all primes whose reciprocals have period length less than 21} ||
| 1 | 3 |
| 2 | 11 |
| 3 | 37 |
| 4 | 101 |
| 5 | 41, 271 |
| 6 | 7, 13 |
| 7 | 239, 4649 |
| 8 | 73, 137 |
| 9 | 333667 |
| 10 | 9091 |
| 11 | 21649, 513239 |
| 12 | 9901 |
| 13 | 53, 79, 265371653 |
| 14 | 909091 |
| 15 | 31, 2906161 |
| 16 | 17, 5882353 |
| 17 | 2071723, 5363222357 |
| 18 | 19, 52579 |
| 19 | 1111111111111111111 |
| 20 | 3541, 27961 |

TABLE 1

discussion to *prime* maximal periods and come up with the following theorem:

Let $q$ be a period length. Then if $q$ is itself a prime greater than 3, the product of all primes whose period lengths are $q$ is the number consisting of $q$ 1's. Conversely, the prime factors (if any )of the number consisting of $q$ 1's are the only primes of period length $q$.

When Gauss was interested in this problem (at the age of 19) he calculated the decimal expansions of the reciprocals of all the primes up to 1000. However, 1000 is not nearly large enough for a stopping place. Table 1 shows all primes with period length less than 21, and it will be seen that in one case it was necessary to reach out into the billions. H. E. Dudeney remarks of the two numbers whose periods are 17 that "their discovery is an exceedingly heavy task." But he wrote this in 1918, and today the task would be made immeasurably easier by suitable machine programming.

The surprising aspect of the table is certainly the small number of primes with short periods: where one might expect hundreds of numbers there are only one, two, or three in each category.

$\bigcirc$

Consider the somewhat strange looking *iterated radical*

$$\sqrt{n + \sqrt{n + \sqrt{n + \sqrt{n + \cdots}}}}$$

Are there any values of $n$ for which this expression converges to a limit that is an *integer*? At first glance an affirmative answer seems exceedingly unlikely. Indeed it looks as if the limit might of necessity be irrational, if it exists at all. Yet, not only is the answer "yes," but the iterated radical, with suitable choice of $n$, can be made to approach any integer whatever greater than 1, and the required $n$ can be prescribed with the greatest of ease.

It is possible to prove that the expression converges to *something*. Let this limit be $x$. Then, because the iteration continues indefinitely, it can likewise be begun at any stage, and

$$x = \sqrt{n + \sqrt{n + \sqrt{n + \sqrt{n + \cdots}}}} = \sqrt{n + x}.$$

Squaring both sides,

$$x^2 = n + x$$

or

$$n = x(x - 1).$$

If $x$ is to be an integer, so is $x - 1$, and we have this simple formula for $n$. For $x = 2$, $n = 2$; for $x = 3$, $n = 6$, etc.

The argument would be fallacious if the expression did not converge in the first place. We omit the proof of this convergence because it involves concepts not otherwise useful to our discussion. What the convergence means, however, is illustrated by the calculation of a few "partial radicals":

$$\sqrt{2} = 1.414\ldots$$

$$\sqrt{2 + \sqrt{2}} = 1.848\ldots$$

$$\sqrt{2 + \sqrt{2 + \sqrt{2}}} = 1.962\ldots$$

$$\sqrt{2 + \sqrt{2 + \sqrt{2 + \sqrt{2}}}} = 1.990\ldots$$

The convergence toward 2 is seen to be fairly rapid.

○

Does $2^{\sqrt{2}}$ have any meaning? We could say, let

$$x = 2^{\sqrt{2}}.$$

Then, remembering how logarithms operate,

$\log x = \sqrt{2} \log 2.$

The right-hand side can be calculated and we can find the corresponding "value" of $x$ from the log tables. However, none of this *explains* what $2^{\sqrt{2}}$ means, or what kind of number it is.

What one must properly do is to recall the meaning of an expression like $2^{3/2}$ in terms of powers and roots:

$2^{3/2} = \sqrt{2^3} = \sqrt{8}$

This is a number with which we are familiar. Now $\sqrt{2}$ does not equal $3/2$, so we have not answered the original question. But in a later chapter we shall show how to find a sequence of rational fractions (like $3/2$) that *approach* $\sqrt{2}$. Each one of these rational fractions $\dfrac{p}{q}$ can be used as an exponent over 2 in a meaningful, familiar way. Then the expression $2^{\sqrt{2}}$ is defined as the limit of $2^{\frac{p}{q}}$ as $\dfrac{p}{q}$ approaches $\sqrt{2}$.

One is not able to tell by looking at them which irrational exponents produce rational quantities. It is easy to show that there exist some expressions $a^b$ where both $a$ and $b$ are irrational, yet the result of raising $a$ to the $b$ power is a rational number:

$\sqrt{2}^{\sqrt{2}}$ is either rational or irrational. If it is rational we need go no farther. If it is irrational, then

$x = \left(\sqrt{2}^{\sqrt{2}}\right)^{\sqrt{2}}$

is a number of the type $a^b$, where both $a$ and $b$ are irrational. But

$x = \left(\sqrt{2}\right)^{\sqrt{2} \cdot \sqrt{2}} = \left(\sqrt{2}\right)^2 = 2,$

is not only rational but integral.

It turns out that the second alternative is the correct one. After long efforts of many mathematicians it was shown that $2^{\sqrt{2}}$ is transcendental.  Therefore

$$2^{\sqrt{2}} = x^{\sqrt{2}} = \left( \left( \sqrt{2}^{\sqrt{2}} \right)^{\sqrt{2}} \right)^{\sqrt{2}} = \left( \sqrt{2}^{\sqrt{2}} \right)^{2}$$

is transcendental, meaning that it is not the solution of any algebraic equation. Therefore neither is $\sqrt{2}^{\sqrt{2}}$, and since $x = 2$, we have an integer expressed as an *irrational* power of a *transcendental* number (see first definition of $x$ above).

# 6

## Diophantine equations

Diophantus of Alexandria (dates unknown, *circa* 3rd century of the Christian era) concerned himself with solutions *in integers* of certain simple algebraic equations. The name Diophantine still attaches to equations for which one wants only integers as solutions. Probably the most familiar example is the Pythagorean equation:

$$x^2 + y^2 = z^2,$$

one solution of which is

$$3^2 + 4^2 = 5^2.$$

It is typical of Diophantine equations that they often have an infinite number of solutions. To "find them all" means to obtain a formula or procedure that turns them out in some systematic fashion.

According to the famous Pythagorean Theorem, solutions of the above equation (such as 3, 4, and 5) can be the sides of a right triangle. Our problem, then, is to find all triples of integers satisfying $x^2 + y^2 = z^2$.

First we note that $6^2 + 8^2 = 10^2$ is not an essentially new solution, because 6, 8, 10 are obtainable from 3, 4, 5 by simply doubling each. Thus an infinite number of others may be obtained as linear multiples of any one, and they are of no particular interest: all such triangles are *similar*. To guarantee that we get new triangles we can consider only the *primitive* solutions, meaning $x$, $y$, $z$ having no factor in common. Note that this means that no two of $x, y, z$ share a factor; for if they did, the third would have to contain it also by the Fundamental Theorem of Arithmetic.

Next we need a small assist from congruence theory. All odd numbers are by definition $\equiv \pm 1 \pmod 4$. Hence all odd squares $\equiv 1 \pmod 4$. Therefore $x$ and $y$ cannot both be odd. For if they were, $z^2$ would be an even square $\equiv 2 \pmod 4$, impossible because all even squares have 4 as a factor (that is, they are $\equiv 0 \pmod 4$).

On the other hand, $x$ and $y$ cannot both be even, for then $z$ would be even and the solution would not be primitive.

Suppose, then, that $x$ is odd and $y$ is even. This means that we can write

$$x^2 + 4u^2 = z^2,$$

with no two of $x, u, z$, having a common factor. Therefore

$$4u^2 = z^2 - x^2$$
$$= (z + x)(z - x).$$

But $x$ and $z$ are odd. Hence both $(z + x)$ and $(z - x)$ are even, say

$$\begin{cases} z + x = 2s \\ z - x = 2r \end{cases}$$

That is,

$$4u^2 = 2s2r$$

or

$$u^2 = rs$$

Now the two bracketed equations yield, on addition, $z = r + s$, and on subtraction, $x = s - r$. If $r$ and $s$ had a common factor, $z$ and $x$ would have it also. But $z$ and $x$ are relatively prime; therefore so are $r$ and $s$. As a consequence, the equation $u^2 = rs$ requires that $r$ and $s$ are *each* perfect squares (neither can pick up a "matching" factor from the other) say, $r = n^2$ and $s = m^2$. Then $u = mn$, and we have finally

$$x = m^2 - n^2$$
$$y = 2mn$$
$$z = m^2 + n^2.$$

What we have done is to show that *if $x, y, z$ are a primitive solution, then* they must have this form. But what are the restrictions on $m$ and $n$? First, $m > n$, so that $x$ will be positive. Second, $m$, $n$ must have no common factor, or it would be shared (twice) by $x$, $y$, and $z$. Third, $m$ and $n$ must not both be odd, for then $x$ and $z$ would share the factor 2. Thus if we let $m$ and $n$ take on all possible arbitrary values within these limitations, we get all primitive solutions. Actually, if we drop out the second and third limitations we still get Pythagorean triangles, but they are not primitive.

**EXAMPLES:**

(1) $m = 3, n = 2$

$\left.\begin{array}{l} x = 9 - 4 = 5 \\ y = 2 \cdot 3 \cdot 2 = 12 \\ z = 9 + 4 = 13 \end{array}\right\}$ Primitive

(2) $m = 4, n = 2$

$\left.\begin{array}{l} x = 16 - 4 = 12 \\ y = 2 \cdot 4 \cdot 2 = 16 \\ z = 16 + 4 = 20 \end{array}\right\}$ Nonprimitive; multiple of 3-4-5 by the square number 4

(3) $m = 5, n = 3$

$\left.\begin{array}{l} x = 25 - 9 = 16 \\ y = 2 \cdot 5 \cdot 3 = 30 \\ z = 25 + 9 = 34 \end{array}\right\}$ Nonprimitive; multiple of 8-15-17 by 2

○

We make use of these findings to prove a geometric theorem: The length of the radius of the circle inscribed in a Pythagorean triangle is always an integer.

This theorem states a fact not at first obvious. There would seem to be insufficient connection between the

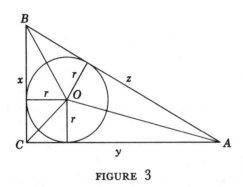

FIGURE 3

radius and the sides to ensure that if the sides are integers, so is the radius. But the proof is easy. Given that $x^2 + y^2 = z^2$, all integers, then $C$ in Figure 3 is a right angle, and the area of the triangle is $\frac{1}{2}xy$. But this area can also be expressed as the sum of the three triangles $BOC$, $COA$, and $AOB$. That is,

$$\frac{1}{2}xy = \frac{1}{2}rx + \frac{1}{2}ry + \frac{1}{2}rz$$
$$= \frac{1}{2}r(x + y + z);$$

solving for $r$,

$$r = \frac{xy}{x + y + z}.$$

But now we know that $x, y, z$ must have the form discovered in the last section. Replacing $x, y, z$ by their expressions in $m$ and $n$ and simplifying,

$$r = \frac{(m^2 - n^2)2mn}{m^2 - n^2 + 2mn + m^2 + n^2} = n(m - n).$$

Not only have we proved that $r$ is an integer; we have also found *which* integer—an extra bonus. In the three examples of the previous section the radii of the inscribed circles are 2, 4, and 6, respectively. Inasmuch as example 2 represented a 3-4-5 triangle with all dimensions multiplied by 4, its $r$ was also multiplied by 4. Therefore the inscribed circle of the 3-4-5 triangle has radius 1—something, possibly, that you never knew before?

O

How many non-collinear points in a plane can be spaced at integral distances each from each? It is clear that an infinite number of points on a straight line can meet the integral demand: simply select all the points whose distances from a fixed point are integers. Then all the differences are certainly integers also. But when we require that not all of the points are to lie on the same straight line the problem takes on interest.

We shall solve it with a set of points all *but one* of which lie on the same straight line. This economical attack squeezes all unnecessary difficulty out of the problem at the outset. The conditions did not demand that the points be scattered all over the plane, but simply that *not all* of them be collinear. By taking just one point not in a line with all the others, we can construct a set of points as large as you please, each of whose distances from all the others is a whole number.

We have found that there are an infinite number of primitive Pythagorean triangles, and we know how to write down as many as we need simply by running pairs of $m$ and $n$ through the formulas. Suppose we are asked to find 7 points fulfilling the conditions of the problem. To do this we use five different primitive Pythagorean triples. Any five will do. To illustrate, we choose five in regular sequence and tabulate the triples thus:

| $m$ | $n$ | $m^2 - n^2,$ | $2mn,$ | $m^2 + n^2$ |
|-----|-----|------------|--------|-----------|
| 2   | 1   | 3,         | 4,     | 5         |
| 3   | 2   | 5,         | 12,    | 13        |
| 4   | 3   | 7,         | 24,    | 25        |
| 5   | 4   | 9,         | 40,    | 41        |
| 6   | 5   | 11,        | 60,    | 61        |

TABLE 2

Now select points in the $(x,y)$-plane according to this plan: One point is the origin, $O$. Another is the point on the $Y$-axis 10395 units from the origin; that is, the point whose coordinates are (0, 10395). Call it $Y$. The number 10395 is chosen because it is equal to $3 \times 5 \times 7 \times 9 \times 11$. For the remaining five, use the points on the $X$-axis whose distances in the positive direction from the origin are, respectively, the $X_i$ of Table 3. It is apparent that $X_1OY$ is a Pythagorean triangle whose hypotenuse is $5 \times 5 \times 7 \times 9 \times 11$; that is, a 3-4-5 triangle each of whose sides has been multiplied by the factor $5 \times 7 \times 9 \times 11$. $X_2OY$ forms a 5-12-13 triangle each of whose sides has been multiplied by the factor $3 \times 7 \times 9 \times 11$, and so on. The part of $X_i$ *not* boxed in Table 3 indicates the factor by which each primitive triple has been magnified. Every hypotenuse $X_i Y$ is an integer. And since this construction can be extended

| $X_1$ | $=$ | 4 | $\times$ | 5 | $\times$ | 7 | $\times$ | 9 | $\times$ | 11 |
| $X_2$ | $=$ | 3 | $\times$ | 12 | $\times$ | 7 | $\times$ | 9 | $\times$ | 11 |
| $X_3$ | $=$ | 3 | $\times$ | 5 | $\times$ | 24 | $\times$ | 9 | $\times$ | 11 |
| $X_4$ | $=$ | 3 | $\times$ | 5 | $\times$ | 7 | $\times$ | 40 | $\times$ | 11 |
| $X_5$ | $=$ | 3 | $\times$ | 5 | $\times$ | 7 | $\times$ | 9 | $\times$ | 60 |

TABLE 3

to any number of triangles, one can produce an arbitrarily large number of points satisfying the conditions of the problem.

○

Having solved so easily the Diophantine equation

$$x^2 + y^2 = z^2,$$

we move on to the general case: what are the solutions in integers of

$$x^n + y^n = z^n$$

for $n > 2$? If you thought the Pythagorean problem too easy, here is a harder one. In fact it is so hard that we advise against your tackling it unless you have a lifetime to spend on the project. It has challenged the finest mathematicians of many generations and has yet to be conquered. That no solutions exist is the famous "last theorem" of Fermat, sometimes called Fermat's *conjecture*. Fermat *said* he had a proof (and he was never dishonest), but he did not publish it. Although some mathematicians now believe that what Fermat thought was a proof may not have been valid, hardly anyone doubts the correctness of the conjecture.

Although we have no general proof covering all values of $n$, Fermat's last theorem has been proved for a large number of specific $n$. The proof for $n = 4$ is one of the simplest. While rather long, its ideas are entirely elementary. We give it here because it illustrates the famous "method of infinite descent."

Let us restrict our attention to primitive solutions, as we did in the Pythagorean case; for if there are no primitive solutions, neither are there any others. To get started we assume the contrary: that some $x^4 + y^4 = Z^4$ can be satisfied by relatively prime $x, y, Z$. If this is so, then of course $Z^4 = z^2$ for some $z$, and so we can say that we are assuming the existence of a solution for

(1)  $x^4 + y^4 = z^2$.

That is,

$$(x^2)^2 + (y^2)^2 = z^2,$$

and $x^2, y^2, z$ form a primitive Pythagorean triple, $x^2$ and $y^2$ not both odd. If we take $y^2$ to be the even one, we know that

$$x^2 = m^2 - n^2$$
$$y^2 = 2mn$$
$$z = m^2 + n^2$$

with $m > n$, $m$ and $n$ relatively prime and not both odd. Now $n$ must be even; for if it were odd we would have

$$x^2 \equiv 0 - 1 \ (\text{mod } 4) \equiv 3 \ (\text{mod } 4),$$

which we decided a short while ago was impossible. That is, $n = 2k$ for some integer $k$ relatively prime to $m$, and

$$y^2 = 2m \cdot 2k.$$

This says, by our old argument of making up a perfect square without any sharing of the factors, that

$$m = r^2, k = s^2, y = 2rs,$$

with $r$, $s$ relatively prime, $r > s$, and $r$ odd. That is,

$$x^2 = r^4 - 4s^4, \qquad \text{or} \qquad x^2 + 4s^4 = r^4.$$

This is a new trio that meets all the conditions of a Pythagorean triple, so here we go again:

$$x = p^2 - q^2$$
$$2s^2 = 2pq$$
$$r^2 = p^2 + q^2$$

$p$, $q$ playing the role of the former $m$, $n$. By the now familiar argument, $pq = s^2$ implies $p = a^2$, $q = b^2$, and at last

$$(2) \qquad r^2 = a^4 + b^4.$$

But this has exactly the form of the original equation

$$(1) \qquad z^2 = x^4 + y^4.$$

We have shown that if (1) is possible, so is (2), which at first glance seems entirely fruitless. But what looks like a complete failure becomes suddenly a rousing success. For how does (2) differ from (1)?

$$z = m^2 + n^2 = r^4 + 4s^4$$

Hence $z > r^4$, and therefore $z > r$, and (2) is a distinctively new solution. Furthermore, a glance at the definition of $r$, *a*, and *b* shows them all to be non-zero: the new solution is not degenerate.

Repeating the whole process will get us another new solution, with some $r_1 < r$; another repetition will yield $r_2 < r_1$; and so on, *indefinitely*. But the sequence

$$r > r_1 > r_2 > \cdots > 0$$

cannot continue indefinitely, because the $r_i$ are all positive integers and there is not an infinite supply of positive integers less than *r*. This is the contradiction of the infinite descent that proves the initial assumption false and gives us the conclusion that

$$x^4 + y^4 = z^4$$

cannot be satisfied in positive integers *x*, *y*, *z*.

O

We present one more geometric picture, this time with the aid of coordinates. Let all the points of the (*x*,*y*) plane *both* of whose coordinates are rational numbers, like (3/2, −49/73), be called "rational points." Between two fractions, however close together, one can always insert another (their average). Hence the rational points constitute a very tight array: to put them "all" in would somehow "fill" the plane. This is what is called an *everywhere dense* set. Nevertheless, even though everywhere dense, the set of points both of whose coordinates are rational does not contain all the points of the plane; for we know that there are points, such as $(\sqrt{2}, \sqrt{3})$, with one or both coordinates irrational. The irrationals form another everywhere dense set, interlarded among the rationals.

Three is one of the many values of $n$ for which Fermat's last theorem has been proved. That is, we know that

$$x^3 + y^3 = z^3$$

has no solution in positive integers. But this is to say that

$$x^3 + y^3 = 1$$

has no solution in *rational* $x$ and $y$ other than $(1,0)$ and $(0,1)$; for if it did, we could clear the fractions to obtain integers

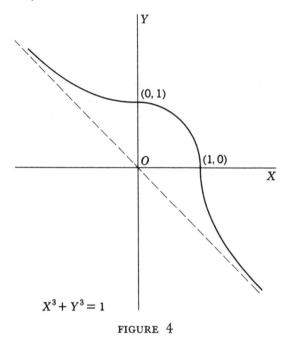

$$X^3 + Y^3 = 1$$

FIGURE 4

satisfying the Fermat equation. The *graph* of an equation means, of course, the curve all of whose points satisfy that equation. Figure 4 indicates what the graph of our last equation looks like. No rational points except $(1,0)$ and $(0,1)$ can satisfy the equation. Therefore, except for these two points, the curve threads its way through the everywhere dense field of rational points without touching a single one of them.

Felix Klein (1849–1925), a great German mathematician and teacher, was the first to notice this phenomenon and was much impressed by it. It turns out to be not so strange as he thought, however, because Cantor has shown that in a certain sense the irrational points are much denser than the rationals: there are more of them. In fact there are so many more of them (a higher order of infinity) that the remarkable thing is rather that a curve wandering through the dense field of points could manage to hit any rational ones at all. There are many examples of curves with Klein's property. We mention one more.

A *transcendental* number is not a solution of any algebraic equation. $\pi$ is a familiar example of such a number and there are infinitely many others. A circle, centered at the origin, with radius $\pi$ (or any other transcendental number) has on it no points both of whose coordinates are rational. For all points of such a circle must satisfy the equation

$$x^2 + y^2 = \pi^2,$$

and $\pi = \sqrt{x^2 + y^2}$ for rational $x$ and $y$ would make $\pi$ merely irrational and not transcendental.

○

In contrast to the Fermat equation, there are infinitely many solutions in integers of

$$x^3 + y^3 + z^3 = w^3,$$

the smallest being

$$3^3 + 4^3 + 5^3 = 6^3.$$

If $w = 870$, $w^3$ can be expressed as the sum of three cubes in no less than nine different ways.

We also have

$$1^3 + 3^3 + 4^3 + 5^3 + 8^3 = 9^3.$$

On the other hand, there is probably no solution to an equation that looks not very different:

$$1^n + 2^n + 3^n + \cdots + k^n = (k + 1)^n$$

except the trivial one, $1 + 2 = 3$. It has been proven that if any other solution exists, it must be for $k > 10^{1.000.000}$.

There is only one solution to the cannon ball problem. One sometimes sees cannon balls piled in a pyramid, perhaps for ornamental purposes, outside an armory. The pile can be formed by setting out a square number of balls for a base, then placing on top of these, for the next layer, a square number one unit less on a side (that is, the next smaller square), and so on up, until the top layer is just one ball. If the eccentric commander of the armory now asks that the total number of balls be also a perfect square, we must solve the Diophantine equation

$$1^2 + 2^2 + 3^2 + \cdots + k^2 = N^2.$$

This differs from the previous problem in that $N \neq k + 1$. It has long been known that the only solution is $k = 24$, $N = 70$, for a total of 4900 cannon balls, but the proof is by no means easy.

○

We attack now another Diophantine equation that has exactly one solution in positive integers:

$$y^2 + 2 = x^3,$$

satisfied only by

$$y = \pm 5, x = 3.$$

In order to point the way toward a proof of this we must first check you out on some high school algebra. Most of the elementary texts on which we were all brought up contain serious blunders, one of which is the unqualified statement that, for all $x$, $a$, and $b$,

$$(x^a)^b = (x^b)^a.$$

It is not always true:

$$((-2)^{1/2})^2 = (\sqrt{-2})^2 = -2$$

but

$$((-2)^2)^{1/2} = \sqrt{4} = 2.$$

To claim that $\sqrt{4}$ is $\pm 2$ begs the question and merely replaces one confusion by another. Both $4^{1/2}$ and $\sqrt{4}$, written without any sign, are always taken to mean 2. If we mean $\pm\sqrt{4}$, we must say so. Another form of the same mistake is the statement that $\sqrt{a} \times \sqrt{b} = \sqrt{ab}$. This is *not* true when $a$ and $b$ are both negative numbers:

$$\sqrt{-9} \times \sqrt{-4} = 3i \times 2i = 6i^2 = -6$$

$$\sqrt{-9} \times \sqrt{-4} \neq \sqrt{36} = 6.$$

We can now give an indication of the proof that

$$y^2 + 2 = x^3$$

has only one solution in positive integers. Let us factor the left-hand side. But, you complain, $y^2 + 2$ is not decomposable. True, it has no real factors. But you can check for yourself, with due observance of the precaution just mentioned, that

$$(y + \sqrt{-2})(y - \sqrt{-2}) = y^2 + 2.$$

We have shifted gears into the realm of objects—call them integers if you like—of the form $a + b\sqrt{-2}$, where $a$ and $b$

are ordinary integers. Let us assume for the moment that the factorization we have just performed has broken $y^2 + 2$ into its *prime* factors in this new kind of integer. If that is so, then by the unique factorization theorem, each of $(y + \sqrt{-2})$ and $(y - \sqrt{-2})$ must be a cube if their product is to be $x^3$. That is, in terms of these integers,

$$y + \sqrt{-2} = (u + v\sqrt{-2})^3$$
$$= u^3 + 3u^2v\sqrt{-2} - 6uv^2 - 2v^3\sqrt{-2}.$$

We must now lean on either your credulity or your knowledge of complex quantities: if two complex numbers are equal, the real part of one must equal the real part of the other, and likewise for the imaginary parts. Equating the coefficients of $\sqrt{-2}$,

$$1 = 3u^2v - 2v^3 = v(3u^2 - 2v^2).$$

Therefore $v$ can only be 1 and $u = \pm 1$. Then, matching the real parts, $y = \pm 1 \mp 6 = \mp 5$.

We said that this was only an outline of a proof. It has two major gaps, and to fill them is well beyond our present scope. The first is the assumption that $(y + \sqrt{-2})$ and $(y - \sqrt{-2})$ are *prime* factors in this kind of integer. It is possible to prove that they are; one cannot decompose them any farther into other factors of this form. You might think that this completes the proof; but there is a second important missing step: how do we know that prime factorization is *unique*, a fact which we need here?

The answer to this question is a long and interesting and very complicated story. It began with the nineteenth-century attempts to prove Fermat's last theorem. Among those who thought for a time that he had proved it was E. E. Kummer (1810–93). He assumed as a matter of course that factorization into primes was always unique, even when the integers were of the form $a \times b\sqrt{-5}$. But this happens to be a field where the Fundamental Theorem of Arithmetic does not hold. Example:

$6 = 2 \times 3$; but also

$6 = (1 + \sqrt{-5})(1 - \sqrt{-5})$.

Each of 2, 3, $1 + \sqrt{-5}$, and $1 - \sqrt{-5}$ can be shown to be prime numbers in this field. So we have the nightmare of two *different* prime factorizations of the number 6. In order to resolve this very grave difficulty Kummer created a new kind of entity that he called an "ideal number." Although he failed to prove Fermat's last theorem, he laid the foundation of a new subject known as the theory of algebraic numbers.

You are now in a position to appreciate that our insistence in Chapter 3 on the importance of unique factorization in ordinary integers was not as vacuous as you might, at the time, have supposed. It so happens that in the field of integers of the form $a + b\sqrt{-2}$, factorization *is* unique; but of course until that can be proven, our proof stands incomplete.

○

In the Diophantine equation of the last section, if we replace the 2 by other constants, we get each time a new problem that has to be handled on its own merits, and these problems all have unrelated answers. Some few have been solved. It is known, for instance, that

$y^2 - 7 = x^3$

has no solutions. On the other hand, there are eight solutions of

$$y^2 - 17 = x^3.$$

When one tries small values of $x$, solutions seem to pop up all over the place.

$$x = -2, -1, 2, 4, 8$$

all yield integral $y$. The next two require greater searching: $x = 43$ and $x = 52$. And the final value that allows $y$ to be an integer is $x = 5234$, for which

$$x^3 = 143,384,152,904.$$

A similar problem that has long defied solution is a complete analysis of the equation

$$a^b - c^d = 1, \quad a, b, c, d, \text{ all different integers } b \text{ and } d \neq 1.$$

A solution is

$$3^2 - 2^3 = 1.$$

Are there any others? How many? What are they? No one knows.

○

There is a vast number of unsolved Diophantine problems, all exceedingly difficult. In 1960 the Polish mathematician Waclaw Sierpinski listed more than forty such problems, most of which have been unfinished business for some time. One of the most innocent in appearance is, can the cube of the sum of three numbers ever equal the product of the numbers? That is, does the equation

$$(x + y + z)^3 = xyz$$

have a solution in integers? Sierpinski presents the problem in three other equivalent forms (see Notes), but makes no headway.

Here are some of the other super brain-teasers that Sierpinski asks us to ponder:

1. Are there any $n$ except 1, 2, and 4 that make $n^n + 1$ a prime? He has shown that if such a prime exists it is greater than $10^{30,000}$.

2. For $n = 4, 5$, and $7, n! + 1$ is a square. Are there any other values of $n$ with this property?

3. How many integral solutions are there of the equation

$$x^3 + y^3 + z^3 = 3,$$

and what are they? The only known ones are the trivial $x = y = z = 1$ and the one consisting of $(4, 4, -5)$ in any order.

4. The same question for

$$x^3 + y^3 + z^3 = 30.$$

Here no solutions are known.

5. Can all numbers be expressed in the form $x^3 + y^3 + 2z^3$? All up to and including 75 can be so expressed. What about 76? The next unknown one is 99.

To these we should also add two very old war horses:

6. *Goldbach's conjecture.* Is every even number expressible as the sum of two primes?

7. Is there an infinite number of twin primes? Twins are pairs whose difference is 2, like $(5, 7)$ or $(29, 31)$.

○

In solving Diophantine problems one must guard against falling into certain formal traps. The identity

$$(a + b)^2 = a^2 + 2ab + b^2$$

might lead one to think that, because all numbers of the form $a^2 + 2ab + b^2$ are squares, any number of, say, the form $a^2 + 3ab + b^2$ could never be a perfect square because it is not an *algebraic* square. But numbers are more versatile than that. Try $a = 7$, $b = 3$ in the expression $a^2 + 3ab + b^2$.

It was once proposed as a theorem that "The product of four consecutive terms of an arithmetic progression of integers plus the fourth power of the common difference is always a perfect square but never a perfect fourth power." If $a$ is the first term of the four and $b$ the common difference, we have

$$a(a + b)(a + 2b)(a + 3b) + b^4,$$

which works out to be the same thing as

$$(a^2 + 3ab + b^2)^2.$$

The truth of the first assertion of the theorem is now evident, but we have already given a counter example showing the second assertion to be incorrect.

O

Inasmuch as nobody buys broken eggs, the following is a Diophantine problem: A farmer sells $p/q$ of his eggs plus $p/q$ of an egg to his first customer, $p/q$ of the remaining eggs plus $p/q$ of an egg to his second customer, and so on until all the eggs have been sold to $n$ customers. Determine necessary and sufficient relations connecting $p$, $q$, and $n$.

The do-it-yourself aspect of such a problem constitutes its main attraction. We leave it to you to discover that $p$ must equal $q - 1$, and that the farmer must start with $q^n - 1$ eggs. (In case total frustration sets in, there is a reference in the Notes.)

# 7
# Number curios

It is assuredly curious that 142857, when multiplied respectively by 2, 3, 4, 5, and 6 yields always a product consisting of a cyclic permutation of the original six digits. But we analyzed this behavior in Chapter 5, answered the question "why," and examined the problem in connection with other allied problems. When such an investigation is possible, the phenomenon in question is much more than a mere curio.

By a "number curio" we mean something perhaps more like the following: Find digits $a$, $b$, $c$, $d$ such that

$$a^b \times c^d = abcd.$$

Joseph de Grazia, who poses this problem, gives an answer

$$2^5 \times 9^2 = 2592$$

and implies that this is the only solution. He states that "there is probably no theoretical equipment that can be used . . . You'll have to try all possible number combinations."

Some number curios appear to be purely fortuitous. Many others are due to peculiarities in the decimal system and do not occur when the same numbers are transformed to a base other than 10. Still others have not very exciting explanations lying near the surface of number theory. A few bearing a superficial appearance of triviality may possibly contain a core of deeper significance that will some day be discovered.

$$\bigcirc$$

It is a well-known joke that the following illegal "cancellation" of 6's yields the right answer:

$$\frac{1\cancel{6}}{\cancel{6}4} = \frac{1}{4}.$$

Are there any other fractions with numerator and denominator each of 2 digits that can be "reduced" in this fashion? It is clear that we have trivial reductions in all cases of the type

$$\frac{4\cancel{4}}{\cancel{4}4} = \frac{4}{4} = 1.$$

In any case, it is necessary that

$$\frac{10x + y}{10y + z} = \frac{x}{z},$$

which says that

$$9xz = y(10x - z).$$

Now if 9 divides $(10x - z)$, we have

$$10x - z \equiv 0 \pmod 9$$

$$10x \equiv z \pmod 9.$$

But

$$10 \equiv 1 \pmod 9$$

so that

$$10x \equiv x \pmod 9$$

for all $x$, and inasmuch as $x$ and $z$ represent single digits, this says that $z = x$ for all $x$, and we have the trivial case. If on the other hand 9 does not divide $(10x - z)$, then 3 must be a factor of $y$. Our search is reduced, but it still requires some trial to discover

$$\frac{26}{65}, \frac{19}{95}, \text{ and } \frac{49}{98}$$

as the only other solutions.

It is more difficult to locate a cancellation like

$$\frac{143\not1\not65}{170\not1\not856} = \frac{1435}{17056}.$$

○

The first few *perfect numbers* are

6, 28, 496, 8128, 33550336, 8589869056.

It was long ago noted that, with the single exception of the first one, if one adds up the digits of any perfect number, then adds up the digits of the resulting sum, and so on, the net result is always 1. Example: $4 + 9 + 6 = 19, 1 + 9 = 10, 1 + 0 = 1$.

We have here a curio within a curio. First, why is the digital sum always 1? And more interesting, why does the rule fail in the case of the first perfect number? Both questions are easily answered.

We recall (page 22) that all even perfect numbers have the form $2^{p-1}(2^p - 1)$, $p$ a prime. All primes are odd numbers (but here you should raise a questioning eyebrow). Therefore $p - 1$ is an even number. Now

$$2 \equiv -1 \pmod 3$$

$$\therefore \quad 2^{p-1} \equiv 1 \pmod 3.$$

This says that $2^{p-1} = 3k + 1$ for some $k$. Multiplying by 2,

$$2^p = 6k + 2$$

$$2^p - 1 = 6k + 1.$$

Hence

$$2^{p-1}(2^p - 1) = (3k + 1)(6k + 1)$$
$$= 18k^2 + 9k + 1 \equiv 1 \pmod 9.$$

That is,

$$2^{p-1}(2^p - 1) - 1 \equiv 0 \pmod 9.$$

But from page 13 we know that any number evenly divisible by 9 has a digital sum of 9. Hence before subtracting 1, our perfect number had a digital sum of 10, and thence 1, which was what we wanted to prove.

Why does this fail for 6? Because it is *not* true that all primes are odd. The single exceptional even prime, 2, yields

$$2^{2-1}(2^2 - 1) = 6$$

and the argument breaks down precisely because 2 is even.

○

It has been proven that the product of a two digit number and its "reversal" (like $57 \times 75$) is never a perfect square except in the obvious case when the two digits are the same (like 55, whose reversal is 55; then, of course, $55 \times 55$ is a perfect square).

The statement does not extend to numbers of more than two digits. For instance,

$$169 \times 961 = 162409 = 403^2$$

and

$$1089 \times 9801 = 10673289 = 3267^2.$$

These examples give rise to the following conjecture: when an integer and its reversal are unequal, their product is never a square except when both are squares. (Note that all the numbers on the left-hand side of both the above equations are, in fact, perfect squares.)

O

Because $10 = 2 \times 5$, it is possible to break some integral powers of 10 into factors containing no zeros. For instance,

$$10^2 = 2^2 \times 5^2 = 4 \times 25$$
$$10^3 = 2^3 \times 5^3 = 8 \times 125.$$

This goes on for a while, but not forever. Up through the exponent 7, the powers of 2 and 5 contain no zeros; but $5^8 = 390625$. Then $2^9$ and $5^9$ are zero-free, but after that the zeros occur with greater frequency. $10^{18}$ and $10^{33}$ are the only other known powers of 10 that can be expressed as the product of two zero-free factors. If there is another one, it is greater than $10^{5000}$.

It is indeed a rather odd curio that

$$8{,}589{,}934{,}592 \times 116{,}415{,}321{,}826{,}934{,}814{,}453{,}125$$
$$= 1{,}000{,}000{,}000{,}000{,}000{,}000{,}000{,}000{,}000{,}000{,}000.$$

O

There is an interesting periodicity in the digits forming the successive powers of 2. If we write down the sequence of these powers,

2, 4, 8, 16, 32, 64, 128, 256, 512, . . .

we observe that the units digit seems to be appearing in the cycle 2-4-8-6. That this cycle continues to recur can be proved, and also that the tens digits appear periodically and so do the hundreds digits, and so on. The length of the period is always $4 \times 5^{n-1}$, where $n$ is the position of the digit counting from the *right*. Thus the length of the period for the unit sdigits $(n = 1)$ is 4, that of the tens digits is 20, etc.

If we look now at the powers of 5 we discern another periodicity.

$$5, 25, 125, 625, 3125, 15625, 78125, 390625, 1953125, \ldots$$

Here, again counting from the right, the period length for the position $n = 1$ is 1; for $n = 2$ it is again 1; for $n = 3$ it is 2; for $n = 4$ it is 4; and in general the period length is $\frac{1}{2} \times 2^{n-1}$ for $n > 1$. Thus whereas

$5^{n-1}$ controls the period length in the powers of 2
$2^{n-1}$ controls the period length in the powers of 5.

This reciprocity is connected with the fact that 2 and 5 are the only proper factors of 10, the base of the number system.

Despite the periodicity in the successive positive integral powers of 2, it is possible to prove that there exist arbitrarily long strings of zeros contained somewhere in the sequence of powers of 2. Actually to find them, however, for large numbers of zeros, becomes an impracticable task. Even for a string of zeros of quite modest length it is necessary to go out to a very large power. The first string of 8 zeros occurs in $2^{14007}$, and the zeros start at the 729th decimal place reading from right to left. No string of 9 zeros occurs in any $2^n$ for $n < 60,000$.

$\bigcirc$

Although $p^2/q^2 = 2$ is never solvable in integers, it is easy to find solutions for

$$\frac{a^2 + b^2}{c^2 + d^2} = 2.$$

For example,

$$\frac{2^2 + 4^2}{1^2 + 3^2} = 2.$$

If the demand is made that the sums of squares be consecutive we have the more spectacular curio

$$\frac{3^2 + 4^2 + 5^2 + 6^2 + 7^2 + 8^2 + 9^2}{1^2 + 2^2 + 3^2 + 4^2 + 5^2 + 6^2 + 7^2} = 2.$$

○

Consider the patterns exhibited in Figure 5.

$$n(n + 1)$$
$$\downarrow$$
$$1 + 2 = 3$$
$$4 + 5 + 6 = 7 + 8$$
$$9 + 10 + 11 + 12 = 13 + 14 + 15$$
$$16 + 17 + 18 + 19 + 20 = 21 + 22 + 23 + 24$$

·   ·   ·   ·   ·   ·   ·   ·   =   ·   ·   ·   ·   ·   ·   ·

$$[2n(n + 1)]^2$$
$$\downarrow$$
$$3^2 + 4^2 = 5^2$$
$$10^2 + 11^2 + 12^2 = 13^2 + 14^2$$
$$21^2 + 22^2 + 23^2 + 24^2 = 25^2 + 26^2 + 27^2$$
$$36^2 + 37^2 + 38^2 + 39^2 + 40^2 = 41^2 + 42^2 + 43^2 + 44^2$$

·   ·   ·   ·   ·   ·   ·   ·   =   ·   ·   ·   ·   ·   ·   ·

FIGURE 5

Both these arrays can be continued indefinitely. Their laws of formation are strikingly similar. The column of figures immediately to the left of the equal signs is the key to both. In the first array the numbers 2, 6, 12, 20, ... are of the form $n(n + 1)$, with $n = 1, 2, 3, 4, ...$ In each equation there are $(n + 1)$ consecutive integers on the left and $n$ on the right. In the second array the numbers 4, 12, 24, 40, ... are just twice $n(n + 1)$, and this time it is the consecutive *squares* adjacent to $[2n(n + 1)]^2$ that are displayed according to the same prescription.

Seeing so much in common to the two arrays we should expect the pattern to be extendable still farther—but it is not. The next array ought to center around values of $[3n(n + 1)]^3$, but we know that it cannot. For if the system were extendable to cubes our first equation, corresponding to $n = 1$, would be

$$5^3 + 6^3 = 7^3,$$

which Fermat's last theorem tells us is impossible.

The interesting point is that this constitutes a very near miss.

$$5^3 + 6^3 = 341$$

and

$$7^3 = 343.$$

If someone had deliberately set out to tease us, he could hardly have done a better job.

What is it that allows the equation

$$x^n + y^n = z^n$$

to have an infinite number of integral solutions for $n = 1$ and $n = 2$, and apparently none for higher $n$? This is the question that has baffled mathematicians for centuries.

# 8

# The primes as leftover scrap

"The primes go on forever"; but their frequency decreases. One of the major achievements of analytic number theory has been the discovery and proof of the Prime Number Theorem, which describes the *asymptotic density* of the primes. The theorem states that as $x$ increases through the sequence of the positive integers, the number of primes less than $x$ tends to become approximately equal to $\dfrac{x}{\log x}$. This means that the approximate or asymptotic density of primes in the vicinity of $x$ is $\dfrac{1}{\log x}$ for large $x$. Stated still another way, the probability that a number of about the size of $x$ is prime is $\dfrac{1}{\log x}$. The logarithm involved here is the natural log, to the base $e$. If you have forgotten, or never were aware of, this kind of logarithm, never mind. We can't prove the prime number theorem anyway; it is far beyond the range of this book. You will have to accept our assurance that the theorem tells us *about* how densely scattered the primes are in any given range. The Prime Number Theorem says that, on the average, the gaps between the successive primes increase in length with increasing $x$, and it approximately quantifies the increase.

It is always possible to exhibit, in the sequence of integers, a prime-free gap as large as desired. If you would like, for instance, to see a sequence of 99 consecutive numbers, all composite, here they are:

$$100! + 2, 100! + 3, 100! + 4, \ldots, 100! + 100.$$

For the first is certainly divisible by 2, since 100! is; the second is divisible by 3; and so on.

Actually this is a very extravagant (though easy) way of obtaining a prime-free gap 99 units long. What we mean by "extravagant" in this case is that we have gone unnecessarily far out in the number sequence to find it. The numbers in the vicinity of 100! are enormous, each containing 158 digits. The Prime Number Theorem tells us that the length of the gaps between primes in that region averages around 360, so that there must certainly be many, many gaps of 99 or more units, long before one reaches 100!.

The first six million primes have been calculated and stored on a magnetic tape. This tape has been analyzed (also by machine) and some interesting statistics obtained. All the differences between successive primes, as well as the frequencies of these differences, were tabulated. The first occurrence of 99 consecutive composite numbers is between the two primes 396733 and 396833: numbers of only 6 digits, not 158. The gaps do not *average* as high as 99, however, until we are in the range where $x$ has about 44 digits. This last bit of information we must extract from the theorem, because no tables go to that length nor anywhere near it. The six millionth prime has only nine digits.

O

Because the primes are distributed so sparsely in the outer reaches of the number sequence, it is natural to ask whether one can form an arithmetic progression (that steps evenly, so to speak, through the number sequence) *none* of whose terms are prime. The answer is immediately yes, of course. An example is the arithmetic progression

10, 15, 20, 25, 30, . . .

all of whose terms are multiples of 5. In fact, if we call the first term $a$ and the common difference $d$, then any arithmetic progression with $a$ and $d$ not relatively prime will have this property; for if $a$ and $d$ have a factor in common, so will every term of the progression.

The less trivial question, then, is the opposite: Is it possible to construct an arithmetic progression none of whose terms are prime, with $a$ and $d$ relatively prime? The answer is no. And it was an outstanding achievement of P. G. L. Dirichlet to prove the stronger theorem that bears his name: *all* such arithmetic progressions contain an *infinite* number of prime terms.

O

One of the aims of analytic number theory has been to refine the Prime Number Theorem to a form that will give close approximation to the exact number of primes less than a given $x$. A famous and useful estimate denoted by $Li(x)$, involving an integral (see Notes), gives a remarkably close approximation in the region where we can get an actual count of the number of primes. Condensed data are summarized in Table 4, where $N$ is the exact number of primes less than $x$. The relative error is seen to be decreasing rather rapidly with increasing $x$. We note, however, that all values of $Li(x)$ in the table are in excess of the correct $N$; and in fact $d$ is not only always positive, it is on the increase. One wonders whether the formula will always produce an

| $x$ | $N$ | $Li(x)$ | $d = Li(x) - N$ | $\dfrac{d}{N} = $ rel. error |
|---|---|---|---|---|
| 1,000 | 168 | 178 | 10 | .060 |
| 10,000 | 1,229 | 1,246 | 17 | .014 |
| 100,000 | 9,592 | 9,630 | 38 | .004 |
| 1,000,000 | 78,498 | 78,628 | 130 | .0017 |
| 10,000,000 | 664,579 | 664,918 | 339 | .0005 |

TABLE 4

excess; that is, does $Li(x)$ approach $N$ asymptotically from above? It would appear so from the table, but this is just a consequence of the fact that our sample is too small. It has been proved by the British mathematician J. E. Littlewood that the sign of $d$ changes, not only once, but infinitely often if we go on indefinitely. How large must $x$ be before $d$ becomes negative for the first time? The answer is not known; but S. Skewes has derived an upper limit. Skewes proves that for some $x$, which he guarantees to be less than $S$, the sign of $d$ will have changed. And $S$, called Skewes' Number, is

$$S = e^{e^{e^{79}}}$$

The incomprehensibly large magnitude of $S$ will be mentioned in the next chapter. It is so big that we will never be able to count primes that far: the applicability of such a result must forever remain purely theoretical.

○

We have noted in Chapter 3 that there is no known formula that turns out the prime numbers. Essentially the only way to find them is by the use of the "sieve" devised by Eratosthenes. First we write down all the numbers 1, 2, 3, . . . as far as we care to go. We then strike out all those we know are *not* primes, and those that are left constitute our table of primes. First every even number goes out, except 2 itself. Then every number divisible by 3 (except 3 itself), that is not already gone; those that were divisible by 2 need not be tested a second time. Now we need not consider numbers divisible by 4, for 4 and all its multiples have already been crossed out as multiples of 2. We go always to the next number that has not been crossed out, leave it in, and cross out all its multiples. The multiples of 5 will therefore go, and the remaining primes will all be contained among the numbers congruent to 1, 3, 7, or 9 modulo 10. These can be arranged in the four columns of Table 5, in which, to

|  1 |  3 |  7 |  9 |
|----|----|----|----|
| 11 | 13 | 17 | 19 |
| 21 | 23 | 27 | 29 |
| 31 | 33 | 37 | 39 |
| 41 | 43 | 47 | 49 |
| 51 | 53 | 57 | 59 |
| 61 | 63 | 67 | 69 |
| 71 | 73 | 77 | 79 |
| 81 | 83 | 87 | 89 |
| 91 | 93 | 97 | 99 |

TABLE 5

save space and time, we stop short of 100. In theory the process could go on indefinitely. The numbers divisible by 3 are crossed out with a single line; the survivors that are divisible by 7 are crossed out with two slant lines; and in this abbreviated range, our search is over. The remaining numbers, plus 2 and 5, are all the 26 primes less than 100. We need not test with 11, 13, etc., because $11 > \sqrt{100}$, and hence if 11 divided some one of the numbers the quotient

| 4 | 7 | 10 | 13 | 16 | . |
|----|----|----|----|----|---|
| 7 | 12 | 17 | 22 | 27 | . |
| 10 | 17 | 24 | 31 | 38 | . |
| 13 | 22 | 31 | 40 | 49 | . |
| 16 | 27 | 38 | 49 | 60 | . |
| . | . | . | . | . | . |

TABLE 6

would be less than 11 and hence would already have been tried as a divisor.

An interesting variant of the sieve that provides an immediate answer on primality is given by Table 6. The first row (and the first column) consists of the arithmetic progression whose $a = 4$ and $d = 3$. The rest of the rows (and columns) now have their $a$'s already fixed. For $d$, use 5 for the second row (column), 7 for the third row (column), and so on. We now have the following tidy criterion: if $x$ is any integer greater than 2, $x$ is prime if and only if $\frac{x-1}{2}$ does *not* appear in the table.

There are various ways to prove this claim. One is, let

$$n = \frac{x-1}{2}$$

then

$$x = 2n + 1.$$

The theorem states that $x$ is composite if and only if $n$ appears in the table. What we are really interested in, then, is not a number in the table but twice its value plus one. Let us make a companion table (7), replacing each entry by twice its value plus one:

|    |    |    |    |     |   |
|----|----|----|----|-----|---|
| 9  | 15 | 21 | 27 | 33  | . |
| 15 | 25 | 35 | 45 | 55  | . |
| 21 | 35 | 49 | 63 | 77  | . |
| 27 | 45 | 63 | 81 | 99  | . |
| 33 | 55 | 77 | 99 | 121 | . |
| .  | .  | .  | .  | .   | . |

TABLE 7

This is rather illuminating. The even numbers are all omitted. The first row (column) contains all odd multiples of 3 except 3 itself, the second row (column) all odd multiples of 5 except 5 itself, and so on. Thus the table lists *all* the odd composite numbers and *only* the odd composite numbers (some more than once, to be sure; the first one to appear three times, for instance, will be $105 = 3 \times 5 \times 7$).

The table efficiently omits the even composites. Therefore, all even primes will slip through this sieve. But we know that there are no even primes except 2 itself, which does indeed slip through, requiring the stipulation in the theorem that $x$ be greater than 2.

The fact that the prime numbers must be found—even *defined*—as the leftovers of a process gives rise to the title of this chapter. The composites can be prescribed in a perfectly systematic writable fashion, as, for example, in the last table. The primes are the ones that remain, almost one might say the bare bones from which the meat has been cut away. But the metaphor is not an altogether bad one: for without the skeleton the structure would be unable to stand. Nevertheless it is this quality of being left behind as the remains of the feast that makes many number theorists doubt whether a positive formula for constructing primes can ever be found.

O

A group of investigators working with Stanislav M. Ulam, the director of the Mathematics Division at Los Alamos Scientific Laboratories, have invented what they call the

| 1 | 3 | 5̸ | 7 | 9 |
|---|---|---|---|---|
| 1̸1 | 13 | 15 | 1̸7 | 1̸9 |
| 21 | 2̸3 | 25 | 2̸7 | 29 |
| 31 | 33 | 3̸5 | 37 | 3̸9 |
| 4̸1 | 43 | 4̸5 | 4̸7 | 49 |
| 51 | 5̸3 | 5̸5 | 5̸7 | 5̸9 |
| 6̸1 | 63 | 6̸5 | 67 | 69 |
| 7̸1 | 73 | 75 | 7̸7 | 79 |
| 8̸1 | 8̸3 | 8̸5 | 87 | 8̸9 |
| 9̸1 | 93 | 9̸5 | 9̸7 | 99 |

TABLE 8

*lucky numbers*, determined also by a sieving process. As with Eratosthenes' sieve, we begin by writing down "all" the natural numbers in order, and again we limit ourselves to the first hundred to illustrate the process. If we leave 1 and strike out every second one after that, we eliminate all the even numbers and are left with Table 8. In Eratosthenes sieve we next struck out every *multiple of 3* because 3 was the next surviving number. Our rule here is different: strike out every *third number* among those remaining. That means that 5 goes, and 11, 17, 23, etc. All such numbers are crossed out by a single slant line. The next surviving number is 7, so we let that stand and cross out every *seventh* remaining one (with two slant lines, to indicate what is happening). Under this axe fall 19, 39, etc. Then every 9th, then every 13th, and so on. The slant lines indicate at what stage in the construction each number was eliminated.

The survivors are called the luckies. In our short table of numbers less than 100, there are 23, of which 10 happen to be prime, 13 composite. Divisibility played no part in their selection, and yet "It turns out that many asymptotic properties of the prime number sequences are shared by the lucky numbers. Thus, for example, their asymptotic density is $\dfrac{1}{\log N}$. The numbers of twin primes and of twin luckies exhibit remarkable similarity up to the integer $n = 100,000$, the range which we have investigated on the machine. The number of adjacent primes differing by 4, or by 6, 8, etc. is, in this range, very similar to the corresponding number of adjacent luckies. It also happens that within the range investigated every even number is a sum of two lucky numbers." (Cf. Goldbach's conjecture, page 82.)

These discoveries present the primes in a new light. That so many properties hitherto thought to be unique to the primes are possessed also by the luckies comes as a distinct surprise. If these properties are consequences only of the fact that the primes are generated by a sieving process and have nothing to do with primality, then the primes have been shorn of some of their distinction. Of course the primes must always play a central role in the development of a large and important body of number theory. We could not get along without them. Nevertheless it seems to be only the randomness of their method of selection that gives them some of the properties connected with their distribution. Ulam suggests that it might be worth investigating the results of other sieving programs. Leftover scrap, perhaps; but it is interesting scrap indeed that organizes itself into two (and possibly more) different and yet strangely similar heaps.

# 9
# Calculating prodigies
# and prodigious calculations

From time to time there have come to public notice young-sters with extraordinary powers of performing mental calculations. As James R. Newman puts it, "To do sums in one's head is no great feat even when the results are correct. But to execute mentally, at high speed, long and complicated numerical calculations—extracting roots, raising to high powers, multiplying and dividing by numbers of ten or twenty digits—is a rare and strange talent."

For some reason not thoroughly understood, many calculating prodigies lose their special powers as they mature. Sometimes, by continuing to cultivate these skills, they have been able to retain and even sharpen them in later life; but more often, as the lumber of everyday existence accumulates in the mental storeroom, there seems to be less room for the youthful attributes, which are gradually crowded out.

Zerah Colburn was an early nineteenth-century mathe-matical whiz-kid, born in a small town in northern Vermont. At the age of nine he was taken by his father to England, where we have a description of his performance before an audience:

He undertook and succeeded in raising the number 8 to the sixteenth power, 281,474,976,710,656. He was then tried as to other numbers, consisting of one figure, all of which he raised as high as the tenth power, with so much facility that the person appointed to take down the results was obliged to enjoin him not to be too rapid. With respect to numbers of two figures, he would raise some of them to the sixth, seventh and eighth power, but not always with equal facility; for the larger the products became the more difficult he found it to proceed. He was asked the square root of 106,929, and before the number could be written down he immediately answered 327. He was then requested to name the cube root of 268,336,125, and with equal facility and promptness he replied 645.

It had been asserted . . . that $4,294,967,297 \ (= 2^{32} + 1)$ was a prime number. Euler detected the error by discovering that it was equal to $641 \times 6,700,417$. The same number was proposed to this child, who found out the factors by the mere operation of his mind.

Zacharias Dase (also Dahse), a German, born in 1824, was perhaps the most remarkable of all the lightning calculators.

Dase was ambitious to make some use of his powers in the service of science. In 1847 he had reckoned out the natural logarithms (7 places) of the numbers from 1 to 1,005,000, and was seeking a publisher. In reckoning on paper he possessed all the accuracy of mental calculation, and added to this an incredible rapidity in doing long problems. In the same year he had completed the calculations for the compensation of the Prussian triangulations. In 1850 the largest hyperbolic table, as regards range, was published by him at Vienna . . . .

In 1850 Dase went to England to earn money by exhibitions of his talents. Much the same is related of his great powers as in Germany; his great obtuseness also occasioned remark. He could not be made to have the least idea of a proposition in Euclid. Of any language but his own he could never master a word.

In 1849 Dase had wished to make tables of factors and prime numbers from the 7th to the 10th million. The Academy of Sciences at Hamburg was ready to grant him support, provided Gauss considered the work useful. [Gauss said yes,] and Dase gave himself up to the execution of the task. Up to his death, in 1861, he had completed the 7th million and also the 8th, with the exception of a small portion. Thus he was able to turn his only mental ability to the service of science, forming a contrast to Colburn and Mondeux, who enjoyed even greater advantages yet failed to yield any results.

He multiplied and divided large numbers in his head, but when the numbers were very large he required considerable time. Schumacher once gave him the numbers 79,532,853 and 93,758,479 to be multiplied. From the moment in which they were given to the moment when he had written down the answer, which he had reckoned out in his head, there elapsed 54 seconds. He multiplied mentally two numbers each of 20 figures in 6 minutes; 40 figures in 40 minutes; and 100 figures in $8\frac{3}{4}$ hours, which last calculation must have made his exhibitions somewhat tiresome to the onlookers. He extracted mentally the square root of a number of 100 figures in 52 minutes.

It is not known how the famous mathematical prodigies of history performed their calculating feats. When questioned they were seldom able to provide lucid explanations. What they apparently did was to devise more or less complicated short cuts which they were then able to remember and use without any written assistance. Whether the ability to do this was inborn or cultivated is impossible to say. One is inclined to believe that, like all mathematical ability, it is some of each. Certainly a powerful memory is of the greatest assistance, and it is true that some people remember numbers better than others.

The most promising young mathematician known to the present writers has some of this ability. One day he wrote on the blackboard

$$e^{\pi \sqrt{163}} = 262,537,412,640,768,743.999,999,999,999,250, \ldots$$

This extraordinary equation is simply an example of what is in all probability a transcendental number, approximated by an integer to an accuracy of 12 decimal places. Our student had picked it up somewhere in his voluminous reading *the previous day* and had *not* copied it down. When questioned as to whether he had just written a random collection of digits to the left of the decimal point, he replied, with an injured glare, "You know I don't do things that way. Those are the *right* numbers." We have never had occasion to check the accuracy of these figures, but we copied them down carefully and we venture to predict that they are indeed correct.

Nearly all of the older books about numbers or mathematical recreations contain accounts of calculating prodigies. The arithmetical wizard was a subject of wonder and admiration to the nineteenth- and early twentieth-century reader. Today there seems to be less general interest in the human calculator, perhaps because his feats have been so far surpassed by his mechanical counterpart. When a Colburn or a Dase multiplied two large numbers together in one-twentieth of the time it would take to do the arithmetic by longhand, people were impressed partly because there was *no other way* to get the answer. But now, when a machine can do the same "problem" in a twenty-thousandth of the time, with virtually no chance of error, the human performance, while just as noteworthy as it always was, fails to make the same impression. It happens that there have been no famous calculating boys born in the twentieth century. When the next one turns up, it is a safe bet that his appearance will cause little stir.

○

In 1853 one William Shanks published the value of pi $(\pi)$ to 607 decimal places. Twenty years later he had extended the work to 707 decimal places, a figure that was to stand for three-quarters of a century. Shanks did his work by longhand (not even a desk calculator existed in those days), and a mere summary of the methods and formulas covers 87 pages. It was considered a great achievement, as indeed in a way it was. In an age when mathematics still meant complicated calculations to most people, here was a concrete result to be admired. Until quite recently a framed copy of the 707-digit decimal could be seen in many a mathematics classroom, a monument to perseverance and what was thought to be accuracy.

In 1949 events took an ironical turn. A (then) modern electronic computer, in three days, turned out 2000 places of $\pi$. When checked against this new determination, Shanks's result was found to be in error after about the 500th decimal place—twenty years' work down the drain. Even prior to 1949 the last 200 numbers of Shanks's determination had been suspect, for an interesting reason. Pi being a transcendental number, its decimal value should show no preponderance of any one digit: each of the 10 digits ought to appear about equally often, in completely random fashion. The expected randomness was not exhibited by Shanks's decimal beyond the 500th place. But of course no one had ever attempted the enormous task of checking the work by hand.

The record of 2000 decimal places for $\pi$ did not stand for long. As computers rapidly increased in size and capacity, $\pi$ was reprogrammed several times after 1949. By 1962 the figure had been pushed to 100,000 decimal places, where it stands today. This determination has been published, and one might quite justifiably ask, what good is it? Since it is no longer a human but merely a mechanical achievement, why bother? One answer has already been suggested. If there were really anything non-random about the distribution of the digits, it would be of great interest to know this because it would imply something unusual and not yet understood about $\pi$. This distribution has in fact shown no discernible pattern, and it was helpful to have a large number of digits to test. As a corollary, we have a second use for the 100,000-digit determination. Mathematicians, both pure and applied, often have need for a *table of random numbers*. How can you select numbers being sure that they are truly random? Here is a ready-made table with 100,000 entries.

O

The numbers that can be manipulated by any human being, however gifted, are of a size well within the average person's comprehension. Those involved in certain parts of number theory, on the other hand, are so vast and fall so far outside the range of comparison with any numbers encountered in "real life" that their size cannot be understood in any ordinary context. In an attempt to describe ultra-large numbers, the American mathematician Edward Kasner invented a new name, the *googol*. Actually, he always insisted that it was his nine-year-old nephew who invented it:

One googol $= 10^{100}$.

A googol, then, is the number consisting of a one followed by a hundred zeros. The word has found its way into many dictionaries. Kasner went a step farther and proposed yet another name:

Googolplex $= 10^{\text{googol}}$.

This is the number one followed, not by just 100 zeros, but by a whole googol of zeros. You may well ask, could we ever meet a number so absurdly large? Yes, we could.

A Fermat number has the form

$$F_n = 2^{2^n} + 1.$$

We have mentioned these briefly in earlier chapters. Partly because of Fermat's conjecture that all such numbers might be prime, the identification of any of their factors has always been of considerable theoretical interest. Fermat's conjecture fails at $n = 5$, for $F_5$ is composite. So many higher Fermat numbers are now known to be composite that the opposite conjecture is in vogue: perhaps *no* Fermat numbers higher than $F_4$ are prime. Consider the probabilities. All $F_n$ from $n = 5$ through $n = 16$ are now known to be composite. $F_{17}$ is so large that we find, on consulting the prime number theorem, that the probability of its being prime is only about $\dfrac{1}{45,000}$ . After that, for higher $F_n$, the probabilities rapidly decrease.

Although partial factorizations of many higher Fermat numbers have recently been discovered with the aid of computers, we summarize only the available information for $n < 17$ in Table 9. The latest of these to succumb to attack by computer were $F_{13}$ and $F_{14}$, found to be composite in 1961 by J. L. Selfridge and Alexander Hurwitz. They remark that $F_{17}$ would be the next to examine, but that "the complete testing of $F_{17}$ would take 128 full weeks of machine time on the IBM 7090."

How big is $F_{17}$? It has 39456 digits. A googol is much

| $n$ | $F_n$ |
|---|---|
| 0, 1, 2, 3, 4 | Prime |
| 5, 6 | Composite and completely factored |
| 7, 8 | Composite but no factors known |
| 9 | Composite; only one prime factor known |
| 10, 11 | Composite; two prime factors known |
| 12 | Composite; three prime factors known |
| 13, 14 | Composite but no factors known |
| 15, 16 | Composite; only one prime factor known |

TABLE 9

smaller. Indeed a googol raised to the 100th power, $(10^{100})^{100}$, has still only about one-quarter as many digits as $F_{17}$. The wonder, of course, is that anything at all can be known about such a number. Yet individual prime factors of many Fermat numbers larger than $F_{17}$ are now known. Needless to say, no one has done any actual long division, even by machine. Relatively simple tests serve to disclose factors of certain very special classes if they exist. The largest Fermat number about which anything is known at present is $F_{1945}$, which has been shown to be divisible by $5 \times 2^{1947} + 1$.

The size of the monster $F_{1945}$ is completely incomprehensible. Of the relatively tiny $F_{36}$, Edouard Lucas wrote, "The ribbon of paper that could contain it would reach around the world." Compare this with the googol, which can easily be written out in full in about two lines of print.

It is of considerable current interest that no factors are known (1965) for $F_7$ or $F_8$, even though it has been known since 1909 that they are composite. They have no small prime factors: the lower limit of the least factor of either has been pushed up to $2^{35} = 34,359,738,368$. $F_7$ has 39 digits; its prime factors therefore can have at most 20 digits. To search for factors in the range of numbers 12 to 20 digits long does not seem an insurmountable task for a large computer today. One guesses that this stubborn fifty-year-old nut is about to be cracked, perhaps even before these pages appear in print.

○

Skewes' number $S$, mentioned in Chapter 8, is to be interpreted as

$$S = e^{[e^{(e^{79})}]} = \text{approximately } 10^{[10^{(10^{34})}]}$$

This number is far bigger than a googolplex, as is indicated in the following table. $S$ is only an upper bound on the size of the first $x$ for which $Li(x) < N$ (see page 96). But if in fact $Li(x) < N$ for the first time "near $S$" in some sense, then it is small wonder that all indications obtainable from among the first six million primes are quite irrelevant.

The only way we can hope to talk meaningfully about the size of these great figures is by comparing them with each other. Table 10 lists a few in order of magnitude.

We close this chapter with a short exercise concerning a long number: How many terminal zeros does 1000! have? Referring to the table we see that 1000! is a number of 2568

| $N$ | Number of digits in $N$ |
|---|---|
| 6-millionth prime | 9 |
| $F_7$ | 39 |
| $F_8$ | 78 |
| Googol | 100 |
| 1000! | 2568 |
| * $2^{11213} - 1$ | 3376 |
| (Googol)$^{100}$ | 10,000 |
| $F_{17}$ | 40,000 (approx.) |
| $F_{36}$ | 20 billion (approx.) |
| Googolplex | $10^{100}$ |
| $F_{1945}$ | $10^{600}$ (approx.) |
| Skewes' number | $10^{10,000,000,000,000,000,000,000,000,000,000,000}$ (approx.) |

* This is the 23rd Mersenne prime, at present (1965) the largest known prime number.

TABLE 10

digits, so we certainly do not want to multiply it out and count the zeros. There must be some better way.

A special notation, known as the greatest integer function, will serve us well in this and similar problems. The symbol $[x]$ means the greatest integer not exceeding $x$. Thus $[\pi] = 3$, $[7/3] = 2$, $[3] = 3$. Inasmuch as $[x]$ seems to be a sort of rough approximation for $x$, it is interesting that we can use it to obtain precise results. It turns out to be just the function we need to solve our problem. In fact, the total number of times any prime factor $p$ appears in the *complete* factorization of $n!$ is

$$\left[\frac{n}{p}\right] + \left[\frac{n}{p^2}\right] + \left[\frac{n}{p^3}\right] + \cdots$$

The three dots here mean that we should continue until we reach zero; that is, until the denominator becomes $> n$.

We might try this on the easy example: How many times does 2 appear in the complete factorization of 9! ?

$$\left[\frac{9}{2}\right] + \left[\frac{9}{2^2}\right] + \left[\frac{9}{2^3}\right] + \left[\frac{9}{2^4}\right] \quad \text{STOP}$$

$$= 4 + 2 + 1 + 0 = 7$$

Check:

$$9! = 1 \times 2 \times 3 \times 4 \times 5 \times 6 \times 7 \times 8 \times 9$$
$$= 1 \times 2 \times 3 \times (2 \times 2) \times 5 \times (2 \times 3) \times 7$$
$$\times (2 \times 2 \times 2) \times (3 \times 3)$$
$$= 1 \times 2^7 \times 3^4 \times 5 \times 7.$$

So we have 7 two's, as predicted. What is the explanation?

The factor 2 occurs once in every even factor of $n!$ written, not in its completely factored form, but in the form $1 \cdot 2 \cdot 3 \cdots n$; that is, in every second factor. It occurs *in addition* once in every fourth factor; in addition to both of those, once again in every 8th factor; and so on. Since 9! has only 9 factors, we were automatically stopped when we divided by 16. If we replace 2 by $p$ we have the general proof.

We note in passing an amusing contrast. Here we are interested only in the whole number of times each divisor is contained in $n$; the remainder is of no consequence, and we discard it. This is the exact opposite of modulo arithmetic, where we cared only about the remainder and not at all about the rest of the quotient.

But we haven't yet answered our original question about the terminal zeros of 1000!, which is now easy. What we need to know is, how many times does the factor 10 appear? But 10 is not a prime, so the formula won't work directly.

However, $10 = 5 \times 2$, and there are far more 2's in 1000! than 5's. Since we have an excess of 2's, we need only ask how many 5's there are:

$$\left[\frac{1000}{5}\right] + \left[\frac{1000}{5^2}\right] + \left[\frac{1000}{5^3}\right] + \left[\frac{1000}{5^4}\right] + \left[\frac{1000}{5^5}\right]$$
$$= 200 + 40 + 8 + 1 + 0 = 249.$$

Hence, 1000! has 249 terminal zeros.

○

Having stretched our minds and our credulity in this chapter by strolling down a few of the lesser lanes and byways, some of which do not properly belong to number theory, we end our digression and return now to the main highway.

# 10

# Continued fractions

We now take a second look at the Euclidean algorithm which we presented in Chapter 3.

On page 29 we confirmed by means of the algorithm that 14 and 45 are *relatively prime*: they have no common factor except 1. Our first operation was to divide 14 into 45. An alternative way to write that step is

$$\frac{45}{14} = 3 + \frac{3}{14}.$$

Next, according to the algorithm, we must divide 3 into 14. In order to put this division into the new pattern, we write

$$\frac{45}{14} = 3 + \frac{1}{\frac{14}{3}}$$

$$= 3 + \frac{1}{4 + \frac{2}{3}}.$$

The next division is 3 by 2, so again we invert:

$$\frac{45}{14} = 3 + \cfrac{1}{4 + \cfrac{1}{\frac{3}{2}}}$$

$$= 3 + \cfrac{1}{4 + \cfrac{1}{1 + \frac{1}{2}}}.$$

Normally we stop here, having arrived at a final numerator of 1. Later we shall have occasion to do the inverting process once more; but if we do, we shall get 1 dividing into 2 evenly, with remainder of zero; and this was also the termination of the Euclidean algorithm whenever the greatest common divisor was 1.

The multiple-decker expression

$$3 + \cfrac{1}{4 + \cfrac{1}{1 + \frac{1}{2}}}$$

is called a *continued fraction expansion* of the number 45/14. We see already its close connection with the Euclidean algorithm.

Suppose now that we *delete* the last fraction, $\frac{1}{2}$, and calculate the value of what is left:

$$3 + \cfrac{1}{4 + \frac{1}{1}} = \frac{16}{5}.$$

This is not equal to 45/14, the number we started with; but then of course we didn't expect it to be. We changed it when we threw away the $\frac{1}{2}$. The two numbers are not the same; but by how much do they differ? Not very much:

$$\frac{45}{14} - \frac{16}{5} = \frac{45 \times 5 - 14 \times 16}{14 \times 5} = \frac{225 - 224}{70} = \frac{1}{70}.$$

Throwing away that $\frac{1}{2}$ made a difference of only 1/70 in the value of the whole expression. But perhaps that was only good luck; we had better try another:

$$\frac{87}{37} = 2 + \frac{13}{37}$$

$$= 2 + \frac{1}{\frac{37}{13}}$$

$$= 2 + \frac{1}{2 + \frac{11}{13}}$$

$$= 2 + \frac{1}{2 + \frac{1}{\frac{13}{11}}}$$

$$= 2 + \frac{1}{2 + \frac{1}{1 + \frac{2}{11}}}$$

$$= 2 + \frac{1}{2 + \frac{1}{1 + \frac{1}{\frac{11}{2}}}}$$

$$= 2 + \frac{1}{2 + \frac{1}{1 + \frac{1}{5 + \frac{1}{2}}}} \cdot$$

Again the last fraction, which we wish to discard, is $\frac{1}{2}$ (it need not have been—that was just happenstance). Deleting it and refiguring the remaining fraction yields

$$2 + \frac{1}{2 + \frac{1}{1 + \frac{1}{5}}} = \frac{40}{17} \cdot$$

As before, we subtract this from the original 87/37 for comparison:

$$\frac{87}{37} - \frac{40}{17} = \frac{87 \times 17 - 37 \times 40}{37 \times 17} = \frac{1479 - 1480}{629}$$

$$= -\frac{1}{629}.$$

The difference is even smaller than before, and this time it happens to be negative. Actually, the important thing is not so much the size of the difference (although we shall use that feature too, in another connection). What is more important at the moment is the fact that the numerator of the final difference fraction is exactly *one* in both examples.

This is not coincidental. Although we omit the proof, it is not hard to show that it always happens provided the numerator and denominator of the original fraction are relatively prime; that is, that the fraction was reduced to "lowest terms" before we started working on it.

From this simple theorem we get several unexpected results.

The linear equation

$$45x - 14y = 1$$

is satisfied by an infinite number of pairs $(x, y)$: all you have to do is select an $x$ and solve for the resulting $y$. But the chances are that they will then *not* both be integers. Are there any Diophantine solutions—that is, pairs of whole numbers $(x, y)$ that satisfy the equation? We have just found one: $x = 5$ and $y = 16$. If you look back a page or two you will find the equation

$$\frac{45 \times 5 - 14 \times 16}{14 \times 5} = \frac{1}{70}.$$

Multiplying through by 70 leaves just what we wanted:

$$45 \times 5 - 14 \times 16 = 1.$$

The *method* of finding the pair (5, 16) was prescribed by the continued fraction development and required no guesswork.

This is not the whole story. Suppose the original equation had been

$$87x - 37y = 1.$$

Our process this time yielded

$$87 \times 17 - 37 \times 40 = -1$$

and this does not solve the equation. But all is not lost; there is a way out. We can extend the continued fraction for 87/37 one more step by replacing the last $\frac{1}{2}$ by

$$\frac{1}{1 + \frac{1}{1}} \cdot$$

Now the final fraction to be discarded is not $\frac{1}{2}$ but $\frac{1}{1}$, and we have to refigure the whole deleted fraction. We leave it to you to do so, trusting that you will come out with

$$2 + \cfrac{1}{2 + \cfrac{1}{1 + \cfrac{1}{5 + \frac{1}{1}}}} = \frac{47}{20} \cdot$$

Finding the difference as usual:

$$\frac{87}{37} - \frac{47}{20} = \frac{87 \times 20 - 37 \times 47}{37 \times 20} = \frac{1740 - 1739}{740} = \frac{1}{740}$$

or

$$87 \times 20 - 37 \times 47 = 1,$$

meaning that the pair $x = 20$, $y = 47$, solves the equation.

Why should anyone *want* to solve such an equation in integers? We hope that by this time you may be in the frame of mind to accept the mountain climber's answer, which is also the mathematician's: "Because it is there." But there are other reasons, if you insist. We can trump up a "real life" situation to which our results apply, although such problems are usually so palpably phony that they are hardly worth manufacturing. A man finds that he can spend all his money on widgets at 87 cents a piece, or he can buy gadgets at 37 cents a piece and have one cent left over. How much money does he have? The cost equation is

$$87W = 37G + 1$$
$$87W - 37G = 1.$$

We have just found that $W = 20$, $G = 47$ satisfy this equation. Therefore he has $87 \times 20 = 1740$ cents, or $17.40.

This is not the *only* solution. Starting with

$$87 \times 20 - 37 \times 47 = 1$$

we can add and subtract the quantity $87 \times 37$ without destroying the equality:

$$87 \times 20 + 87 \times 37 - 37 \times 47 - 87 \times 37 = 1$$
$$87(20 + 37) - 37(47 + 87) = 1$$
$$87 \times 57 - 37 \times 134 = 1.$$

Therefore he might have had $49.59, enough to buy 57 widgets or 134 gadgets. By adding and subtracting integral multiples of $87 \times 37$ we can find any number of solutions.

The geometric picture is not devoid of interest. The point (20, 47), having both its coordinates integers, is called a lattice point of the plane. It is a point of intersection of a horizontal and a vertical line of the coordinate grid, or lattice. The equation

$$87x - 37y = 1$$

represents a straight line passing through the lattice point (20, 47). It does not pass through another such point for some distance; the next is (57, 134). But what we have proven above shows that if a line whose equation has the form $ax - by = 1$ passes through one lattice point, it must eventually pass through an infinite number of them.

If we had more time we should next discuss the equation

$$ax - by = c,$$

where $a$ and $b$ are not relatively prime. If their greatest common divisor is $c$, we divide through by it and this case reduces to the previous one. If it is not $c$ but some other number, say $d$, one can soon show that

$$ax - by = c$$

is solvable in integers if and only if $c$ is evenly divisible by $d$.

○

It is clear that no irrational number can have a finite (terminating) continued fraction expansion; for we know that we can always "condense" any finite continued fraction until we have telescoped it all back into a single rational number $p/q$. Therefore if the irrationals have continued fraction expansions (and indeed they do), these must be nonterminating.

To find the continued fraction expansion for $\sqrt{2}$, we begin by adding and subtracting the largest integer less than $\sqrt{2}$:

$$\sqrt{2} = 1 + \sqrt{2} - 1.$$

We had good luck inverting before, so we try it again:

$$\sqrt{2} = 1 + \cfrac{1}{\cfrac{1}{\sqrt{2}-1}}.$$

Now the denominator of the last fraction, $\dfrac{1}{\sqrt{2}-1}$, can be

rationalized by multiplying by $\dfrac{\sqrt{2}+1}{\sqrt{2}+1}$ :

$$\sqrt{2} = 1 + \cfrac{1}{\cfrac{\sqrt{2}+1}{(\sqrt{2}-1)(\sqrt{2}+1)}}$$

$$= 1 + \cfrac{1}{\cfrac{\sqrt{2}+1}{2-1}}$$

or

$$\sqrt{2} = 1 + \cfrac{1}{1+\sqrt{2}}.$$

Now comes the critical maneuver, the key to the whole process. We can replace the final $\sqrt{2}$ by the entire right-hand side, for that is also equal to the square root of 2:

$$\sqrt{2} = 1 + \cfrac{1}{1+\left(1+\cfrac{1}{1+\sqrt{2}}\right)}$$

$$= 1 + \cfrac{1}{2+\cfrac{1}{1+\sqrt{2}}}.$$

But there seems to be nothing to prevent us from repeating this performance. So we take our courage in both hands and say that IF the process actually converges to anything, it must converge to $\sqrt{2}$. In fact, it does.

$$\sqrt{2} = 1 + \cfrac{1}{2 + \cfrac{1}{2 + \cfrac{1}{2 + \cdots}}}$$

where the three dots mean that the fraction does not terminate.

To check (not prove) that our continued fraction is approaching $\sqrt{2}$, let us evaluate successive *convergents*, the partial fractions obtained by starting from the left. The first is $C_1 = 1$, if we delete all the rest. The next few:

$$C_2 = 1 + \cfrac{1}{2} = \cfrac{3}{2}$$

$$C_3 = 1 + \cfrac{1}{2 + \cfrac{1}{2}} = \cfrac{7}{5}$$

$$C_4 = 1 + \cfrac{1}{2 + \cfrac{1}{2 + \cfrac{1}{2}}} = \cfrac{17}{12}$$

and so on. If you had actually to do a large number of these, you would soon detect a pattern.

$$\frac{1}{1}, \frac{3}{2}, \frac{7}{5}, \frac{17}{12}, \cdots$$

Each new denominator is the sum of numerator and denominator of the preceding fraction. $7 + 5 = 12$; thus the next denominator in the sequence should be $17 + 12 = 29$. To get the new numerator, add the new denominator to the previous one: in this case, $29 + 12$, for $41$. So the next fraction is $\frac{41}{29}$, the next after that $\frac{99}{70}$, etc.

Now how do these compare with $\sqrt{2}$? The best way to find out is to square them and compare the squares with 2. Here are the squares:

$$\frac{1}{1}, \frac{9}{4}, \frac{49}{25}, \frac{289}{144}, \frac{1681}{841}, \frac{9801}{4900}, \cdots$$

The successive *differences* between these fractions and the number 2 are

$$-\frac{1}{1}, +\frac{1}{4}, -\frac{1}{25}, +\frac{1}{144}, -\frac{1}{841}, +\frac{1}{4900}, \cdots$$

Note that the signs alternate, the numerators are all 1's, and the denominators increase rapidly. We met all these characteristics in the previous section in connection with the truncated finite continued fraction.

Each fraction of the sequence is a better approximation to $\sqrt{2}$ than its predecessor, and it can be shown that there are none "in between," so to speak. We observed in Chapter 5 that the irrationality of $\sqrt{2}$ is equivalent to the fact that no perfect square is twice another. What we are finding here is all those pairs of squares that are *almost* in the ratio of 2 to 1. If the convergents are denoted by $y/x$, they all satisfy

$$\frac{y^2 \pm 1}{x^2} = 2$$

or

$$y^2 - 2x^2 = \pm 1.$$

The continued fraction development for $\sqrt{2}$ gives us all the integral solutions of a quadratic Diophantine equation, just as a continued fraction gave us all the solutions of a linear Diophantine equation.

These solutions can be strikingly displayed on a graph. Imagine the first quadrant of the $(x,y)$ plane to be a board with pins stuck perpendicularly into all the lattice points. We now mark in some distinctive way, say with a red-headed pin, each lattice point whose coordinates constitute a

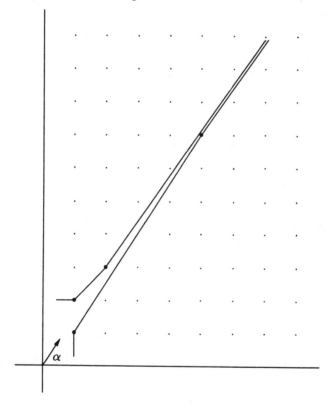

FIGURE 6

solution of the equation: $(1, 1)$, $(2, 3)$, $(5, 7)$, $(12, 17)$, etc. If we now stretch a thread from the pin $(1, 2)$ to the pin $(2, 3)$, thence to $(12, 17)$, and so on, using every lattice point associated with a quotient *greater* than $\sqrt{2}$, the thread will form a polygonal line as shown in Figure 6.

A second thread can be passed around (1, 1), (5, 7), and all the other red-headed pins associated with quotients *less than* $\sqrt{2}$. These two threads will bound a corridor *devoid of lattice points*. Let the arrow in the diagram point in the direction of the exact irrational value of $\sqrt{2}$ (meaning $\tan \alpha = \sqrt{2}$); then if you were to look in that direction from the origin you would, theoretically, have a clear view, unobstructed by pins, all the way "to infinity." Furthermore, except for the terminal pin at (1, 2) which we used only to get things started, the taut threads will touch *only* the red-headed pins and *all* of them.

○

We are now able to answer the question asked in Chapter 2: What perfect squares are also triangular numbers? We found at that time that triangular numbers have the form $(n^2 + n)/2$. When is this equal to a square?

$$\frac{n^2 + n}{2} = m^2.$$

Clearing, multiplying through by 4, and adding 1 to both sides,

$$4n^2 + 4n + 1 = 8m^2 + 1$$

or

$$(2n + 1)^2 = 2(2m)^2 + 1.$$

This is solvable in integers whenever

$$y^2 - 2x^2 = 1$$

is solvable in integers, with $y = 2n + 1$, $x = 2m$. We now know that these solutions are exactly every alternate convergent of the continued fraction for $\sqrt{2}$: (2, 3), (12, 17), (70, 99), . . . Inasmuch as $m = x/2$, the required squares are $1^2$, $6^2$, $35^2$, . . .

○

All irrational numbers of the form $\sqrt{a^2 + 1}$ can be developed just as we handled $\sqrt{2}$; and in fact the general solution, of which $\sqrt{2}$ is a particular case, is

$$\sqrt{a^2 + 1} = a + \cfrac{1}{2a + \cfrac{1}{2a + \cfrac{1}{2a + \cdots}}}.$$

But the procedure must be revised for all numbers not falling into this category. For instance, $\sqrt{3}$, is not a number of the form $\sqrt{a^2 + 1}$. Its expansion, which we derive in the Notes, is

$$\sqrt{3} = 1 + \cfrac{1}{1 + \cfrac{1}{2 + \cfrac{1}{1 + \cfrac{1}{2 + \cdots}}}}.$$

Having solved the equation

$$y^2 - 2x^2 = \pm 1$$

with such signal success by means of $\sqrt{2}$, we would expect that $\sqrt{3}$ might serve us equally faithfully for

$$y^2 - 3x^2 = \pm 1.$$

The successive convergents are

$$1, 2, \frac{5}{3}, \frac{7}{4}, \frac{19}{11}, \frac{26}{15}, \cdots$$

whose squares are

$$1, 4, \frac{25}{9}, \frac{49}{16}, \frac{361}{121}, \frac{676}{225}, \cdots$$

How do these compare with 3? Subtracting 3 from each we get

$$-\frac{2}{1}, +\frac{1}{1}, -\frac{2}{9}, +\frac{1}{16}, -\frac{2}{121}, +\frac{1}{225}, \cdots$$

Come now, that can't be right. What are those 2's doing in the numerators? There must be some mistake. We calculate again, and there is no mistake. This is indeed a jolt. If the convergents are denoted by $y/x$, then every *second* convergent is such that

$$y^2 - 3x^2 = 1,$$

but the others satisfy

$$y^2 - 3x^2 = -2,$$

an equation that we did not ask for but which has intruded itself gratuitously. There is nothing we can do about this; it is part of life in the number jungle.

It turns out that the equation

$$y^2 - Nx^2 = 1,$$

known as Pell's equation, has solutions in integers whenever $N$ is not a perfect square, but that

$$y^2 - Nx^2 = -1$$

has solutions for $N = 2$ but not for $N = 3$. Exactly which $N$'s provide solutions is part of a long and interesting story that cannot be completed here.

Everything in this chapter so far has involved at most *quadratic* irrationals—*square* roots. Although $\sqrt{3}$ was not as kind to us as $\sqrt{2}$, nevertheless its continued fraction was periodic. This is a perfectly general state of affairs: all quadratic irrationals have periodic continued fraction developments. *Cubic* irrationals, on the other hand, are quite another matter, about which very little is known.

○

The great and prolific mathematician Leonhard Euler pointed out that if a convergent series has the form

$$c_1 + c_1 c_2 + c_1 c_2 c_3 + \cdots$$

then it is equivalent to the continued fraction

$$\cfrac{c_1}{1 - \cfrac{c_2}{1 + c_2 - \cfrac{c_3}{1 + c_3 - \cdots}}}$$

You can verify this by multiplying back the first few convergents.

Now there is a well-known series for the angle whose tangent is $x$:

$$\arctan x = x - \frac{x^3}{3} + \frac{x^5}{5} - \frac{x^7}{7} + \cdots$$

$$= x + x\left(-\frac{x^2}{3}\right) + x\left(-\frac{x^2}{3}\right)\left(-\frac{3x^2}{5}\right) + \cdots$$

$$= \cfrac{x}{1 + \cfrac{x^2}{3 - x^2 + \cfrac{9x^2}{5 - 3x^2 + \cfrac{25x^2}{7 - 5x^2 + \cdots}}}}$$

In the special case where $x = 1$ we get on the left the angle whose tangent is 1, or 45°. This is $\frac{\pi}{4}$ (radians, if you like), so that finally,

$$\frac{\pi}{4} = \cfrac{1}{1 + \cfrac{1^2}{2 + \cfrac{3^2}{2 + \cfrac{5^2}{2 + \cdots}}}}$$

That the transcendental number $\pi$, which is the root of no algebraic equation (not quadratic or cubic or any other degree), should have a systematic continued fraction expansion is something of a surprise. We note, however, that the expansion is not what the technician calls *simple*; this means that the partial numerators are not all 1's. It is possible to write the beginning of a simple continued fraction for $\pi$, and a large number of convergents of the simple fraction development have in fact been found; but the denominators do not appear to follow any predictable pattern.

The number $e$, the base of the natural logarithm, is defined by the convergent series

$$e = 1 + \frac{1}{1!} + \frac{1}{2!} + \frac{1}{3!} + \cdots$$

Like $\pi$ it is transcendental, but unlike $\pi$ its simple continued fraction expansion shows a pattern:

$$e = 2 + \cfrac{1}{1 + \cfrac{1}{2 + \cfrac{1}{1 + \cfrac{1}{1 + \cfrac{1}{4 + \cfrac{1}{1 + \cfrac{1}{1 + \cfrac{1}{6 \cdots}}}}}}}}$$

Although far afield from number theory, this opens up an intriguing and very difficult question: Is $e$ in some sense a "simpler" number than $\pi$? D. Shanks and J. W. Wrench, Jr., tell us that the machine time required to compute the value of $e$ to 100,000 decimal places was less than one-third of the time required for the corresponding calculation of the value of $\pi$. They then remark, in typical Shanksian style, "One would hope for a theoretical approach $\cdots$—a theory of the '*depth*' of numbers—but no such theory now exists. One can guess that $e$ is not as 'deep' as $\pi$, but try and prove it!"

# I I

# Fibonacci numbers

In Chapter 10 we did not consider the simplest possible continued fraction, consisting entirely of 1's:

$$1 + \cfrac{1}{1 + \cfrac{1}{1 + \cfrac{1}{1 + \cdots}}}$$

Letting the value of this fraction be $x$, we have

$$x = 1 + \frac{1}{x}$$

or

$$x^2 - x - 1 = 0$$

This quadratic equation has two solutions:

$$x = \frac{1 \pm \sqrt{5}}{2},$$

but only the one with the plus sign is admissible because the minus sign would yield a negative value of the continued fraction, not possible. Hence we have

$$x = \frac{1 + \sqrt{5}}{2} = 1.618\ldots$$

This number is sometimes designated by the Greek letter $\phi$ (although there is no uniformity; it has been given various names in the literature).

The convergents of the continued fraction are

$$\frac{1}{1}, \frac{2}{1}, \frac{3}{2}, \frac{5}{3}, \frac{8}{5}, \frac{13}{8}, \frac{21}{13}, \frac{34}{21}, \ldots$$

In comparison with the successive convergents of $\sqrt{2}$ on page 123, or in fact any sequence of convergents in the last chapter, these seem to be changing rather slowly. Because delicate approximation to an irrational number by means of successive rational fractions requires large numerators and denominators, one can guess that this continued fraction converges comparatively slowly to $\phi$. For instance, 99/70, the sixth convergent of $\sqrt{2}$, differs from $\sqrt{2}$ by .000072; but 13/8, the sixth convergent of $\phi$, differs from $\phi$ by .0070, showing an error nearly 100 times as large. In fact it can be shown that of all possible continued fractions, the convergence of the $\phi$-fraction is the slowest.

The law of formation of the sequence of convergents is easily discernible. We immediately observe that the numerators are all the same as the denominators "moved over one." The denominators are

1, 1, 2, 3, 5, 8, 13, 21, 34, 55, 89, 144, 233, . . .

This famous sequence is named after Leonardo da Pisa, a thirteenth-century mathematician called *Fibonacci*. It has been intensively studied and is still the subject of much investigation by number theorists.

The Fibonacci numbers are such that, after the first two,

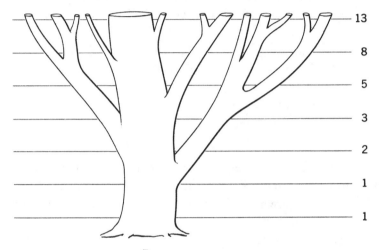

FIGURE 7. THE FIBONACCI TREE

every number in the sequence equals the sum of the two previous numbers:

$$F_n = F_{n-1} + F_{n-2}.$$

The sequence is sometimes encountered in nature. Suppose a tree grows according to the following not unrealistic formula. Each old branch (including the trunk) puts out one new branch per year; each new branch grows through the next year without branching, after which it qualifies as an old branch. The growth is represented schematically in Figure 7. The number of branches after $n$ years is $F_n$.

○

The Fibonacci numbers have a large number of easily derivable properties and an even larger number of only slightly more difficult ones. Among the simplest is: the sum of the first $n$ Fibonacci numbers is $F_{n+2} - 1$. We have

$$F_1 = F_3 - F_2$$
$$F_2 = F_4 - F_3$$
$$\vdots$$
$$\vdots$$
$$F_{n-1} = F_{n+1} - F_n$$
$$F_n = F_{n+2} - F_{n+1}$$

If we add up all these equations, we get on the left-hand side the sum of the first $n$ $F$'s. On the right we have cancellation of the type known to the trade as *telescoping*, leaving only $F_{n+2} - F_2$. But $F_2 = 1$.

Another property is

$$F_{n+1}^2 = F_n F_{n+2} + (-1)^n.$$

The last term means that the sign in front of the final 1 alternates. For instance, we have

$$n = 6: \ F_7^2 = F_6 F_8 + 1; \ 13^2 = 8 \times 21 + 1$$
$$n = 7: \ F_8^2 = F_7 F_9 - 1; \ 21^2 = 13 \times 34 - 1$$

and so on. This is easy to prove by induction. It is true for $n = 1$: $1^2 = 1 \times 2 - 1$. Now suppose it is true for $n = k$:

$$F_{k+1}^2 = F_k F_{k+2} + (-1)^k$$

Add the quantity $F_{k+1}F_{k+2}$ to both sides:

$$F_{k+1}^2 + F_{k+1}F_{k+2} = F_k F_{k+2} + F_{k+1}F_{k+2} + (-1)^k$$

Factoring,

$$F_{k+1}(F_{k+1} + F_{k+2}) = F_{k+2}(F_k + F_{k+1}) + (-1)^k.$$

But remembering that, by definition, $F_k + F_{k+1} = F_{k+2}$, we replace each expression in parenthesis accordingly, to get

$$F_{k+1}F_{k+3} = F_{k+2}^2 + (-1)^k$$

This says

$$F_{k+2}^2 = F_{k+1}F_{k+3} - 1(-1)^k$$

or

$$F_{k+2}^2 = F_{k+1}F_{k+3} + (-1)^{k+1}.$$

Thus we have shown that if the property is true for $n = k$, then it is also true for $n = k + 1$, and the induction is complete.

○

There is an unexpected connection between the binomial coefficients and the Fibonacci numbers. Figure 8 represents

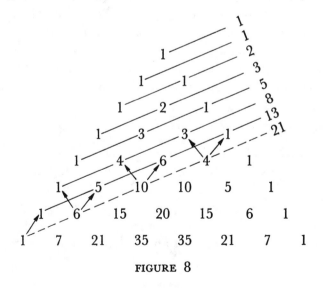

FIGURE 8

the Pascal Triangle of page 48 with embellishments. The diagonal lines are called the *rising diagonals* of the triangle. You may be surprised (we hope) to find that the *sums* of the numbers on the rising diagonals form the Fibonacci sequence. Blaise Pascal died in 1662, and his triangle was doubtless known to others even before he made it famous. Yet more than 200 years were to pass before anyone troubled to add up the diagonals. In 1876 a compatriot of Pascal's, the number theorist Edouard Lucas, discovered the relation that now seems so obvious. As usual, it was a question of looking at an old familiar scene from a new angle—in this case an angle of $22\frac{1}{2}°$.

Why the sums along these diagonals are the Fibonacci numbers is "not hard to see," although we shall not pretend that this is a proof. The sequence starts correctly with two 1's. Therefore what we have to convince ourselves is that the numbers on any diagonal, say for instance along the dotted line, add up to the sum of all the numbers on the two preceding diagonals. According to the original law of formation of the triangle (see page 48), each number along the dotted diagonal is the sum of the two numbers to which the arrows point; they exactly account for all the numbers on the two preceding diagonals and no others.

○

Our next Fibonacci property has a built-in geometric proof. If we start with two squares each 1 unit on a side, adjoin to them a 2 × 2 square, then add to that picture a 3 × 3 square, and so on, we obtain a rectangle like that shown in Figure 9. If we happen to stop at the 8 × 8 square, we have the area of an 8 × 13 rectangle expressed

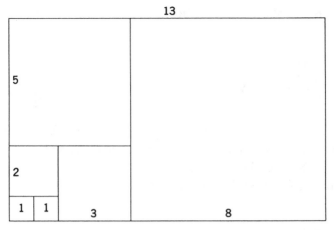

FIGURE 9

in the form

$$1^2 + 1^2 + 2^2 + 3^2 + 5^2 + 8^2 = 8 \times 13.$$

We can carry the construction to any stage, so that the general property can be written

$$F_1{}^2 + F_2{}^2 + F_3{}^2 + \cdots + F_n{}^2 = F_n F_{n+1}.$$

We now ask another geometric question. Where is the point that divides a line into two segments such that the longer part is the mean proportional between the whole line and the shorter part? If we start with a line of length 1 (Figure 10), we require the length $x$ such that

$$\frac{1}{x} = \frac{x}{1 - x}.$$

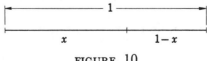

FIGURE 10

This is the quadratic equation

$$x^2 + x - 1 = 0$$

whose solutions are

$$x = \frac{-1 \pm \sqrt{5}}{2}.$$

We select the plus to get a positive value of $x$. (By the way, what is the geometric meaning of the negative value here?) What, then, is the ratio of the length of the whole line, 1, to this value of $x$? It is expressed by

$$\frac{1}{x} = \frac{2}{-1 + \sqrt{5}}$$

$$= \frac{2}{-1 + \sqrt{5}} \times \frac{1 + \sqrt{5}}{1 + \sqrt{5}}$$

$$= \frac{2(1 + \sqrt{5})}{-1 + 5}$$

$$= \frac{1 + \sqrt{5}}{2}.$$

But this is precisely $\phi$. The ancients called such a division a Golden Section.

Suppose we are given the large rectangle of Figure 11, whose sides are in the Golden Ratio of $\phi:1$. Incidentally, the psychologists tell us that this is supposed to be in some sense the most pleasing shape for a rectangle. It is approximately the shape chosen for picture postcards, etc. If we now take away the $1 \times 1$ square, the remaining rectangle has its sides in the ratio of $\dfrac{1}{\phi - 1}$. But we know that

$$\phi^2 - \phi - 1 = 0$$

is the defining equation for $\phi$. That is,

$$\phi^2 - \phi = 1$$

$$\phi(\phi - 1) = 1$$

$$\phi = \frac{1}{\phi - 1}.$$

This says that the new rectangle on the left is exactly similar to the original rectangle. We can therefore repeat the

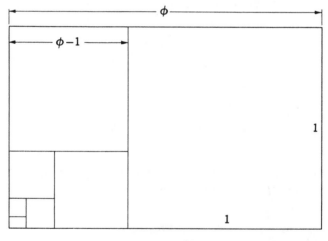

FIGURE 11

process, obtaining Figure 11, which looks very much like Figure 9. The two diagrams differ considerably in the lower left-hand corners; but the larger parts are nearly alike. And if we were to increase both figures by building on squares, the proportions of the two pictures would become *increasingly more nearly equal.* The ratio of length to width of *every* oblong of Figure 11 is $\phi$; the proportions of the oblongs of Figure 9, being the successive convergents, approach $\phi$ as the sides increase in length. We noted earlier that 13/8 differs from $\phi$ by only about .007.

The Golden Ratio appears in the 5-pointed star or Mystic Pentagram of old (Figure 12). Each isosceles triangle forming a "point" of the star has 72° in the base angles and

FIGURE 12. THE PENTAGRAM

36° at the apex. This means that in Figure 13, triangles *ABC* and *CBD* are similar, and hence

$$\frac{AB}{BC} = \frac{BC}{BD} \quad \text{(corresponding sides)}$$

$$\frac{AB}{AD} = \frac{AD}{BD} \quad \text{(because } AD = BC\text{)}$$

which says that *D* divides *AB* in a golden section. Also the ratio of the side of the isosceles triangle forming the point of the star, to its base, is $\phi$.

In 1963 a group of enthusiasts who felt that there was much interesting work still to be done in this field founded the Fibonacci Association and began publication of a quarterly journal devoted principally to researches on Fibonacci numbers. In its first two years of existence the journal published some 600 pages covering the results of investigations in this particular field.

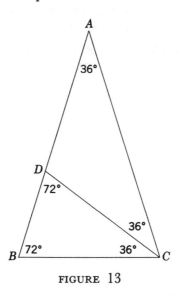

FIGURE 13

The Fibonacci Association has adopted for its symbol a device consisting of nested pentagrams.

○

There are those who feel that anything like a Fibonacci Association has backed itself into too small a corner of the intellectual landscape. Nevertheless some nationally known mathematicians are counted among its members and have contributed to its journal. We repeat the often stated principle that it is hazardous indeed to pass judgment on what does or does not constitute valuable mathematics.

Besides, what if a study is not of earth-shaking importance? If it stimulates the imagination and whets the appetite for more, is not that enough? Do we dare to hope that this book has done as much for you? Have we cast a little light on what was formerly dim, so that you now wish you knew more about some of these things?

The path is endless, but many rewards are offered along the way. One could do worse than follow the gleam of numbers.

# NOTES

The number refers to the
corresponding page of the text

1. Modern investigators tend to discount the earlier claims concerning the counting ability of birds and animals. Today there seems to be little experimental evidence to indicate that they have any number sense at all.

2. Ten raised to the zero power is *defined* as equal to 1. But it is a reasonable definition. For any positive integral exponents $m \neq n$, we have

$$\frac{10^m}{10^n} = 10^{m-n}$$

If this rule is to hold even if $m = n$, then $10^0 = 1$.

6. See "Ethiopian multiplication," C. S. Ogilvy, *Pentagon*, Fall 1950, p. 17.

8. There are many forms of Nim. In one of the standard ways to play the two-man game, three piles of chips are thrown at random on the table. Each player in turn removes one or more chips from *only one* pile. He must remove at least one, and he may take the whole pile if he wants to. The player forced to remove the last chip loses.

Briefly, the winning technique consists in confronting your opponent with a *balanced situation*, thus forcing him to unbalance it, whereupon you balance it again until you reach the end situation where it is his move and the piles show either 3-2-1 or 2-2-0 or 1-1-1. Any of these positions is hopeless for him.

A balanced situation is one in which all powers of 2 (digits in the binary system) are paired. Suppose the initial piles contain 28, 31, and 17 chips:

28: $16 + 8 + 4$
31: $⓰ + 8 + 4 + ② + 1$
17: $16 \qquad\qquad\quad + 1$

Everything pairs automatically except the 16's and the 2. Therefore if it is your move you should take 18 chips from the 31 pile, and your opponent will be doomed. You are not, of course, obligated to take all of a power of 2. For example, it is now his move with the circled numbers gone. If he elects to take all of the 17 pile, how do you balance? By removing 15 from the 28 pile. Try it!

This explanation is given in No. 30 of *Ingenious Mathematical Problems and Methods*, L. A. Graham (Dover, New York, 1959).

12. From *Carl Friedrich Gauss, Titan of Science*, G. Waldo Dunnington (Hafner, New York, 1955), p. 12.

14. A prime number is any integer greater than 1 that has only itself and 1 as exact divisors. By conventional agreement, 1 is not a prime. 2 and 47 are examples of primes; 20 is composite, being factorable into $2 \times 2 \times 5$.

In number theory one deals most frequently with the positive integers. If we are sometimes careless and use simply the word *number* when we mean positive integer, we trust that the context will leave no doubt about our meaning.

14. The formula

$$n^2 - 79n + 1601$$

produces primes for all values of $n$ up to and including 79, but fails when $n = 80$:

$$80^2 - 79 \times 80 + 1601 = 1681 = 41^2.$$

17. The theorems at the end of this section and some others were independently rediscovered and proved by Keith Backman, now (1965) at the University of Chicago, while he was a high school junior.

18. For the sum of two consecutive triangular numbers, add algebraically:

$$\frac{n^2 + n}{2} + \frac{(n + 1)^2 + (n + 1)}{2}.$$

20. For those interested in pursuing further the properties of triangular numbers, "On the representation of numbers as the sum of triangular numbers," U. V. Satyanarayana, gives a few references (*Mathematical Gazette*, vol. 45, 1961, p. 40).

20. Kepler: see, for instance, *Mathematics, a Cultural Approach*, Morris Kline (Addison-Wesley, Reading, Mass., 1962), p. 255.

20. The Mersenne quotation is reproduced in *Mathematical Recreations*, Maurice Kraitchik (Dover, New York, 1953), p. 72.

20. The number $10^{25}$ is from an unpublished paper by Dr. Mariano Garcia of the University of Puerto Rico, relayed to us by Professor John L. Selfridge of Pennsylvania State University. It probably represents the correct 1965 state of knowledge concerning a lower

bound for the least odd perfect number, which is continually being raised. (A much greater lower bound was mistakenly deduced and published in Tokyo in 1956 and later retracted by its Japanese author. *Mathematical Reviews*, vol. 20 (1959), no. 3095.)

20. "The springboard for an entire book," *Solved and Unsolved Problems in Number Theory*, Daniel Shanks (Spartan, Washington, D.C., 1962).

22. Marin Mersenne (1588–1648).

22. Leonhard Euler (1707–83), one of the most prolific and gifted mathematicians of all time.

22. The 23 values of $p$ for which it is known that $2^p - 1$ is prime, together with the story of the discovery of the last three, is given in "Three new Mersenne primes and a statistical theory," Donald B. Gillies, *Mathematics of Computation*, vol. 18 (1964), p. 93.

23. The problem on abundant numbers was published (actually for the second time) in 1962 in *Tomorrow's Math, Unsolved Problems for the Amateur*, C. S. Ogilvy (Oxford University Press, New York).

23. Abundant Numbers, Thomas R. Parkin and Leon J. Lander (Aerospace Corp., Calif., July 1964).

26. It is important that the lemmas do not depend upon the uniqueness of factorization, for that is what we are trying to prove. The whole question of "how careful can you be" in constructing a proof is a sticky one. We shall frequently in this book not be completely rigorous in our proofs; but we shall make some attempt to indicate when we are really proving something and when we are doing no such thing. Mathematicians will note a missing element in this "proof."

30. What is an unbiased coin? Obviously, one that is just as likely to fall heads as tails. But if we talk about two

events being "equally likely" we mean that their probabilities are equal; so we are going around in circles. A
purely mathematical definition of probability that does
not require the *a priori* concept of equal likelihood has
yet to be devised.

Some people say, "Toss a coin a large number of
times. Divide the number of heads that appear by the
total number of tosses. As the number of tosses increases,
this quotient approaches a fixed limit, called the *probability* of obtaining a head on one toss." The trouble with
this empirical definition of probability is that the
experimenter anticipates that the probability will turn
out to be $\frac{1}{2}$. In fact he knows in advance that it must—
because he knows in advance what he *wants* probability
to mean! If a very large number of tosses should produce systematically more heads than tails, would he
conclude that the probability of a head is in general
greater than $\frac{1}{2}$? On the contrary, he would conclude
that he had been tossing a biased coin.

32. $\dfrac{\pi^2}{6}$ . Any standard text in advanced calculus will do;
for instance, *Advanced Calculus*, Angus Taylor (Ginn &
Co., Boston, 1955), Ex. 3, p. 717.

45. *Theorem:* If $a \equiv b \pmod{m}$, and if $ac \equiv d \pmod{m}$,
then $bc \equiv d \pmod{m}$.

*Proof:* (everything is understood to be mod $m$, which
we omit for brevity) $a \equiv b$ means $a = b + km$.

Now if $ac \equiv d$, then, replacing $a$ by its equal,

$$(b + km)c \equiv d$$

$$\text{or } bc + kmc \equiv d$$

that is, $bc - d \equiv -kmc \equiv 0$

$$\therefore \ bc \equiv d.$$

48. Pascal's Triangle: the third diagonal column consists of all the *triangular numbers*. Why?

49. The binomial coefficient $\dfrac{n!}{k!\,(n-k)!}$ is often abbreviated $\dbinom{n}{k}$. We have mentioned a few of the elementary properties of the $\dbinom{n}{k}$; but not all arbitrary relations can be fulfilled. For instance, it has been shown that

$$\binom{2n}{n} = \binom{2a}{a}\binom{2b}{b}$$

has no solution in integers. "Notes on number theory, V," Leo Moser, *Canadian Mathematical Bulletin*, vol. 6 (1963), p. 167.

50. Composite exponent cannot divide all its binomial coefficients. For let $k$ be the smallest prime factor of $n$. The $(k+1)$st term of the expansion is

$$j = \frac{(n)(n-1)\cdots(n-k+1)}{k!}$$

If $n$ divided $j$, then

$$\frac{(n-1)\cdots(n-k+1)}{k!}$$

would be an integer, which is not possible, because none of the factors of the numerator is divisible by $k$.

51. Even if the converse of Fermat's Theorem were true it would not provide a *workable* criterion for primality; one cannot routinely deal with numbers of 100 or perhaps many more digits. We were able to handle easily the case $n = 341$ only because we knew in advance what its factors were.

51. To show that the sum of all the numbers in the $n$'th row of the Pascal Triangle $= 2^n$, simply expand $2^n$ in the form $(1+1)^n$.

52. There are only four other values of $n < 2000$, besides 341, that are composite and yet divide $2^n - 2$:

   561,    1387,    1729,    1905.

   *History of the Theory of Numbers*, L. E. Dickson (Stechert, New York, 1934), vol. 1, p. 94. Dickson is the standard reference for everything that happened in number theory up to about 1915.

57. The "maximal period" primes less than 100 are 7, 17, 19, 23, 29, 47, 59, 61, and 97. In "A note on primitive roots," *Scripta Mathematica*, vol. 26 (1963), p. 117, D. H. Lehmer gives some interesting information on the frequency of maximal period primes and discusses two freak cases.

59. "Primes and recurring decimals," by R. E. Green, is a very elementary treatment of some periodicity properties of repeating decimals; *Mathematical Gazette*, vol. 47 (1963), p. 25.

59. The answer to question (a) has been known for only about half a century. Information on (b) was greatly augmented in 1963 when it was ascertained that the only strings of 1's less than 110 units long that are prime numbers are the numbers consisting of 2, 19, and 23 1's respectively. "Some miscellaneous factorizations," John Brillhart, *Mathematics of Computation*, vol. 17 (1963), p. 447.

61. Gauss's table of reciprocals of primes can be found in his *Werke*, Band 2, p. 412.

61. For easily readable papers on period lengths of repeating decimals the interested reader is referred to the *American Mathematical Monthly*, vol. 56 (1949), p. 87; vol. 62 (1955), p. 484; vol. 66 (1959), p. 797.

62. For more on the iterated radical, see problem E-874, *American Mathematical Monthly*, vol. 57 (1950), p. 186.

67. Brother U. Alfred, F.S.C., has a discussion of "Consecutive integers whose sum of squares is a perfect square," a special class of Pythagorean triples, in *Mathematics Magazine*, vol. 37 (1964), p. 19.

69. The problem is E-1528, *American Mathematical Monthly*, vol. 70 (1963), p. 440.

72. Fermat's last theorem has been proved for *all* positive integral exponents up to 4000. "Proof of Fermat's last theorem for all prime exponents less than 4002," John L. Selfridge, C. A. Nichol, and H. S. Vandiver, *Proceedings of the National Academy of Science*, vol. 41 (1955), p. 970.

72. Fermat's last theorem for $n = 4$: a closely allied result is that the area of a Pythagorean triangle can never be a perfect square. See Dickson, vol. 2, p. 615.

76. For the complete solution of the first equation in this section, see a shorter book by the same L. E. Dickson, *Introduction to the Theory of Numbers* (U. of Chicago Press, 1946), p. 58.

77. "Nine different ways," Leon Bankoff, in *American Mathematical Monthly*, vol. 64 (1957), p. 507, E-1249.
    For a fourth power as the sum of four fourth powers, see *Mathematics Magazine*, vol. 37 (1964), p. 322.

77. First equation: Dickson, vol. 2, p. 682. Second equation: Leo Moser in *Scripta Mathematica*, vol. 19 (1953), p. 84.

77. The cannon ball problem: Dickson, vol. II, p. 25, gives several references.

80. There is a brief treatment of non-unique factorization in *Elementary Introduction to Number Theory*," Calvin T. Long (D. C. Heath, Boston, 1965), p. 32. For more, see *The Theory of Algebraic Numbers*, Carus Monograph No. 9, by Harry Pollard (Math. Assoc. of Amer., 1950).

81. Solution of $y^2 - 17 = x^3$: L. J. Mordell in *Proceedings of the London Mathematical Society* (2), vol. 13 (1913), p. 60. Louis J. Mordell is a living contradiction of the oft-stated proposition that all creative mathematics is produced by young men. Professor Mordell's many contributions to number theory cover a span of fifty years and still continue to appear. In 1964 one of the present authors was privileged to hear him deliver at the age of 76 an energetic and inspiring lecture.

81. $a^b - c^d = 1$ is called Catalan's problem. J. W. S. Cassels has a large list of references on the problem in *Proceedings of the Cambridge Philosophical Society*, vol. 56 (1960), p. 97. See also K. Inkeri "On Catalan's problem," *Acta Arithmetica*, vol. 9 (1964), p. 285, and Seppo Hyyro (in Finnish), *Mathematical Reviews*, vol. 28 (1964), p. 13, no. 62.

81. Sierpinski: "On some unsolved problems of arithmetics," *Scripta Mathematica*, vol. 25 (1960), p. 125. Could there be something about the field of number theory that encourages longevity and extraordinary continuation of mathematical productivity? Waclaw Sierpinski provides an even more spectacular example than Mordell. "In 1962 at 80 he traveled through the United States with almost youthful vigor" to give a series of lectures at various universities. He then returned to his native Warsaw to work (*Scripta Mathematica*, vol. 27 (1964), p. 105). Even as we go to press he is still publishing original mathematics at 83—an unheard of achievement (*Mathematical Reviews*, vol. 29 (1965), p. 426, no. 2215). The reviewer in this instance comments, "The author [Sierpinski] proves several interesting theorems . . . "—which in this most concise and austere journal of abstracts is high praise.

82. No. 1: *L'Enseignement mathématique*, vol. 4 (1958), p. 211.

   Nos. 3 and 4: Problem E-1555, *American Mathematical Monthly*, vol. 70 (1963), p. 896, contributes some but not much information.

   No. 7: It is well known that $\Sigma 1/p$ diverges if $p$ runs over all the primes. In 1921 Viggo Brun proved that $\Sigma 1/q$ *converges* if $q$ runs over only the twin primes. Does this indicate enough difference between the density of the primes and the density of the twin primes to suggest that the number of twins is only finite? The question has not been answered. Brun's proof (not easy) is given in Edmund Landau (not easy either), *Elementary Number Theory* (Chelsea, New York, 1958), p. 94.

83. "It was once proposed . . . , " *American Mathematical Monthly*, vol. 57 (1950), p. 186, E-876..

83. "A farmer sells his eggs," *Mathematics Magazine*, vol. 26 (1953), p. 164.

85. Joseph de Grazia, *Math Is Fun* (Emerson Books, New York, 1954), p. 143.

87. The last pseudo-cancellation is from *Mathematical Games and Pastimes*, A. P. Domoryad (Macmillan, New York, 1964), p. 35. Somewhere along the way, in the translation and republication from the Russian, this book has acquired a vast number of misprints, sometimes as many as five or six on a page. One of the errors supplies us with an unintentional problem. It is incorrectly stated (p. 35) that

$$\frac{4251935345}{91819355185} = \frac{425345}{9185185}.$$

If it were so, it would be an example of a "correct" pseudo-cancellation. Presumably the author originally had this one right. Where is the error?

88. "Conjecture on reversals," *American Mathematical Monthly*, vol. 64 (1957), p. 434, E-1243.

89. $10^{5000}$: "Non-zero factors of $10^n$," Rudolph Ondrejka, *Recreational Mathematics Magazine*, no. 6 (1961), p. 59.

90. See E-942, *American Mathematical Monthly*, vol. 58 (1951), p. 418. In fact, in the sequence of positive integral powers of *any* digit, the sequence of digits in any (decimal) place is periodic. For there are only $10^k$ ways to write a $k$-digit number, including unlimited use of zeros. If $d$ is any digit, and $n$ is an exponent such that $d^n$ has $k$ digits, then the same right-hand set of $k$ digits will recur, and in the same order, at some $m \leq n + 10^k$, thus establishing periodicity for the $k$'th digit from the right.

*Challenging Mathematical Problems with Elementary Solutions*, Yaglom and Yaglom (Holden-Day, San Francisco, 1964), poses as problem 90 the following: "What is the probability that the first digit of $2^n$ is a 1?" Here $2^n$ is meant to be any randomly selected integral power of 2. We paraphrase the solution given therein.

In the sequence of positive integral powers of 2, there are three numbers of one digit each, namely, 2, 4, and 8; several of length 2 digits, 3 digits, etc. There are always some, $x$ digits long, for any integer $x > 1$. Furthermore, the first and *only* the first one of them begins with the digit 1, for *each x*. These two facts we leave to you to prove; the proofs are not difficult, depending only on the fact that successive powers of 2 are connected by a simple *doubling*.

Now suppose $2^n$ is a number of $x$ digits, and $q(n)$ stands for the number of powers of 2 in the sequence (of powers of 2) up to and including $2^n$, that begin with the digit 1. What we have just said means that $q(n) = x - 1$ (not $x$, because in the first decade none of the one-digit numbers 2, 4, 8 qualify).

Since $2^n$ has $x$ digits, $x - 1$ is the characteristic of $\log_{10} 2^n$. That is, $\log_{10} 2^n = x - 1 + \alpha$, where $\alpha$ is a quantity between 0 and 1. Hence

$$x - 1 = \log_{10} 2^n - \alpha.$$

But $x - 1$ is $q(n)$, and our question is, what happens to the probability $\dfrac{q(n)}{n}$ as $n$ increases? Specifically,

$$\lim_{n \to \infty} \frac{q(n)}{n} = \lim_{n \to \infty} \frac{\log_{10} 2^n - \alpha}{n}$$

$$= \lim_{n \to \infty} \frac{n \log_{10} 2 - \alpha}{n}$$

$$= \lim_{n \to \infty} \left[ \log_{10} 2 - \frac{\alpha}{n} \right]$$

$$= \log_{10} 2 = .30103 \cdots$$

It is rare that logarithms to any base other than $e$ appear in combinatorial problems. The base 10 owes its presence here to the fact that the problem depends on the decimal notation.

90. "Arbitrarily long strings of zeros," *American Mathematical Monthly*, vol. 70 (1963), p. 1101, E-1565. Also "The first power of 2 with 8 consecutive zeros," E. and U. Karst, *Mathematics of Computation*, vol. 18 (1964), p. 646.

91. The "more spectacular curio" is from *Scripta Mathematica*, vol. 21 (1955), p. 201, where one also finds others of its ilk.

91. From a different formula, J. S. Vigder has found

$$4^2 + 5^2 + 6^2 + \cdots + 37^2 + 38^2$$

$$= 39^2 + 40^2 + \cdots + 47^2 + 48^2.$$

*Mathematics Magazine*, vol. 38 (1965), p. 42.

92. For those who seek more number curios, the last chapter of Dickson, vol. 1, has many; the earlier *Scripta Mathematica* volumes are full of them; and *Mathematics Magazine* sometimes carries a few, e.g. Norman Anning's "Surprises," vol. 36 (1963), p. 80.

93. The prime number theorem was proved independently in 1895 by Jacques Hadamard and C. de la Vallee-Poussin.

94. "Statistics on the first six million prime numbers," F. Gruenberger and G. Armerding (The RAND Corp., Santa Monica, Calif., 1961).

94. See also "On maximal gaps between successive primes," Daniel Shanks, *Mathematics of Computation*, vol. 18 (1964), p. 646.

95.

$$Li(x) = \int_0^x \frac{du}{\log u} \text{ (natural log, of course)}$$

96. Skewes's paper appears in *London Mathematical Society Journal*, vol. 8 (1933), p. 277. There is a further note on the problem in *A Mathematician's Miscellany*, John E. Littlewood (Methuen & Co., London, 1960), p. 113.

EXCURSIONS IN NUMBER THEORY

99. This version of the sieve is mentioned in *Scripta Mathematica*, vol. 8 (1941), p. 164, where it is credited to a young Indian mathematician, S. P. Sundaram.

100. *A Collection of Mathematical Problems*, Stanislav M. Ulam (Interscience Publishers, New York, 1960), p. 120.

103. The James R. Newman quotation is from his four-volume collection, *The World of Mathematics* (Simon & Schuster, New York, 1956), vol. 1, p. 465.

103. Completely reliable information on the calculating prodigies is no longer available, if it ever was. The long quotations are from "Arithmetic prodigies," E. W. Scripture, *American Journal of Psychology*, vol. 4 (1891), p. 1. It is the basic reference on this rather archaic subject; yet as early as 1907 its accuracy was questioned by Frank D. Mitchell in a rambling 82-page dissertation in the same journal (vol. 18, p. 61).

Accuracy is a quality that does not improve with age, and one is never sure who is responsible for the introduction of errors. Scripture quotes from a letter purporting to have been written by the famous 17th-century English mathematician John Wallis, as follows:

"December 22d, 1669.—In a dark night, in bed, without pen, ink or paper or anything equivalent, I did by memory extract the square root of 30000,00000,-00000,00000,00000,00000,00000,00000, which I found to be 1,77205,08075,68077,29353, *feré*, and did the next day commit it to writing."

As every schoolboy knows, $\sqrt{3} = 1.732 \ldots$, and Wallis would never have knowingly written $1.772 \ldots$, especially when he was trying to show off a little. The 14th digit in the result attributed to Wallis is also in error. Yet Scripture copied the number as it appeared in an 1879 source (*Spectator*, vol. 52, p. 11), and it can be traced from there to the *Classical Journal*, vol. 7 (1815), p. 179. All sources contain the same errors, but we are still 145 years away from Wallis when the trail ends.

107. For the William Shanks dates, see *A Budget of Paradoxes*, Augustus de Morgan (Open Court, Chicago, 1915), vol. 2, p. 63.

107. The following table shows the mechanical progress since 1949 in the calculation of $\pi$ by electronic computer.

| Author | Machine | Date | Decimal places | Time (hours) |
|--------|---------|------|----------------|--------------|
| Reitwiesner | ENIAC | 1949 | 2000 | 70 |
| Nicholson & Jeenel | NORC | 1954 | 3000 | 0.2 |
| Felton | Pegasus | 1958 | 10,000 | 33 |
| Genuys | IBM 704 | 1958 | 10,000 | 1.7 |
| Unpublished (Paris) | IBM 704 | 1959 | 16,000 | 4.3 |
| Gerard | IBM 7090 | 1961 | 20,000 | 0.7 |
| Shanks & Wrench | IBM 7090 | 1961 | 100,000 | 9 |

The table is made from data given in "Calculation of $\pi$ to 100,000 decimals," Daniel Shanks and John W. Wrench, Jr., *Mathematics of Computation*, vol. 16 (1962), p. 76, wherein further details and references may be found. In the same article the authors speculate about future extensions of the work:

"Can $\pi$ be computed to 1,000,000 decimals with the computers of today? From the remarks in the first section we see that the program which we have described would require times in the order of *months*. But since the memory of a 7090 is too small, by a factor of ten, a modified program, which writes and reads partial results, would take longer still. One would really want a computer 100 times as fast, 100 times as reliable, and with a memory 10 times as large. No such machine now exists. There are, of course, many other formulas ... and other programming devices are also possible, but it seems unlikely that any such modification can lead to more than a rather small improvement."

109. "Googol," *Mathematics and the Imagination*, Edward Kasner and James R. Newman (Simon & Schuster, New York, 1940), p. 20.

109. The prime number theorem says that the probability that a randomly chosen number of about the size of $F_{17}$ is prime is about $1/90,000$. But we know *a priori* that $F_{17}$ belongs in the smaller collection of all odd numbers, which doubles the probability.

110. For a more complete listing of the known data on $F_n$, see "A report on primes of the form $k \cdot 2^n + 1$ and on factors of Fermat numbers," Raphael M. Robinson, *Proceedings of the American Mathematical Society*, vol. 9 (1958), p. 673; also "New factors of Fermat numbers," Claude P. Wrathall, *Mathematics of Computation*, vol. 18 (1964), p. 324.

110. In the *Notices* of the American Mathematical Society, vol. 8 (1961), p. 60, Alexander Hurwitz and John L. Selfridge announced that $F_{14}$ is composite. Their remark on $F_{17}$ (in quotes) is from "Fermat and Mersenne numbers," *Mathematics of Computation*, vol. 18 (1964), p. 146.

110. Edouard Lucas, *Théorie des nombres* (1891), vol. 1, p. 51. The world-girdling ribbon would contain about 15 digits per inch.

111. $2^{35}$: "Some miscellaneous factorizations," John Brill-hart, *Mathematics of Computation*, vol. 17 (1963), p. 447.

112. The 23rd Mersenne prime: "Three new Mersenne primes . . . ," Donald B. Gillies, *Mathematics of Computation*, vol. 18 (1964), p. 93.

113. On 1000!, see also E-1180, *American Mathematical Monthly*, vol. 63 (1956), p. 189.

120. P. P. Kelisky has pointed out that, among all the lattice points on the line, the first Diophantine solution is always that lattice point nearest the origin. "Concerning the Euclidean algorithm," *Fibonacci Quarterly*, vol. 3 (1965) p. 219.

121. Inasmuch as the slope is rational, the result about lattice points is immediate; but we prefer not to assume any knowledge of analytic geometry.

121. For a geometric derivation of the continued fraction expansion of $\sqrt{2}$, see *Through the Mathescope*, C. S. Ogilvy (Oxford University Press, New York, 1956), p. 16.

122. Carl D. Olds has written an excellent elementary presentation of continued fractions for the monograph project of the School Mathematics Study Group, intended for outside reading by intelligent and ambitious high school students; *Continued Fractions* (Random House, New York, 1963). The last statement in the section ending on (our) page 121 is the main result in Olds's Chapter 2.

126. Squares that are also triangular numbers: "Ten mathematical refreshments," Dewey Duncan, *The Mathematics Teacher*, vol. 58 (1965), p. 102. Sierpinski does this an altogether different way starting on p. 20 of his book, *Pythagorean Triangles*.

127. We develop the continued fraction for $\sqrt{a^2 + 2}$, of which $\sqrt{3} = \sqrt{1 + 2}$ is a special case.

$$\sqrt{a^2 + 2} = a + \sqrt{a^2 + 2} - a$$

$$= a + \cfrac{1}{\cfrac{1}{\sqrt{a^2 + 2} - a}}.$$

Now "rationalize the denominator" of the last part:

$$\frac{1}{\sqrt{a^2 + 2} - a} = \frac{1}{\sqrt{a^2 + 2} - a} \cdot \frac{\sqrt{a^2 + 2} + a}{\sqrt{a^2 + 2} + a}$$

$$= \frac{\sqrt{a^2 + 2} + a}{2} = \frac{2a + \sqrt{a^2 + 2} - a}{2}$$

$$= a + \frac{\sqrt{a^2 + 2} - a}{2} = a + \cfrac{1}{\cfrac{2}{\sqrt{a^2 + 2} - a}}$$

So far, we have arrived at

$$\sqrt{a^2 + 2} = a + \cfrac{1}{a + \cfrac{1}{\cfrac{2}{\sqrt{a^2 + 2} - a}}}.$$

Now $\dfrac{2}{\sqrt{a^2 + 2} - a} = 2\left(\dfrac{1}{\sqrt{a^2 + 2} - a}\right)$, and we just

finished developing the fraction in parentheses. Writing its value into the main expression, we get

$$\sqrt{a^2 + 2} = a + \cfrac{1}{a + \cfrac{1}{2a + \cfrac{1}{\cfrac{1}{\sqrt{a^2 + 2} - a}}}}$$

And this is where we came in: the show repeats again, and again, . . .

$$\sqrt{a^2 + 2} = a + \cfrac{1}{a + \cfrac{1}{2a + \cfrac{1}{a + \cfrac{1}{2a + \cdots}}}}.$$

We observe that $a = 1$ yields the result for $\sqrt{3}$ given on page 127.

129. The question of period length of the repeating decimal of a rational fraction has an analogue here in the period of the repeating part of the continued fraction expansion of a quadratic irrational. What can we say about these periods? Very little. Olds's book (see note to page 122) has an interesting table of the periodic part of the expansion of $\sqrt{N}$ up to $N = 40$, examination of which raises immediate unanswered questions. Why is the period of $\sqrt{31}$ longer than any other? What is it that $\sqrt{13}$ and $\sqrt{29}$ have in common that they should be the only ones with period length 5? Why are there no periods of length 3? (See Olds, p. 116.)

Professor Olds wrote (1965) in a letter to the authors, "I do not think much is known about the length of the period of continued fractions. I encouraged some people at Lockheed Aero. Lab. to compute the periods of quadratic irrationals in the hope that something might be discovered. But all I have are two thick volumes of numbers . . . . The scope of the program used to compute this table is amply demonstrated by . . . the continued fraction expansion for $\sqrt{1,000,099}$. This expansion has a period of 2174 terms . . . . The smallest positive integer $x$ satisfying

$$x^2 - 1,000,099y^2 = 1$$

is 1118 digits long."

Thus while it is true that Pell's equation has always a solution, to find it may be quite another matter!

131. The quotation is from the Shanks and Wrench paper cited in the note to p. 100.

138. Edouard Lucas, *Nouveau Correspondance mathématique*, vol. 2 (1876), p. 74. Lucas also invented and studied another Fibonacci-like sequence that now bears his name. The Lucas numbers satisfy the same recursion relation as the Fibonacci numbers, but have starting values $L_1 = 1$, $L_2 = 3$.

139. Fig. 9. It seems hardly necessary to add anything to a geometric proof that stands so firmly on its own feet; but for those who like analytic methods we give the easy inductive proof.

$1 + 1 = 1 \times 2$ for the first step. Then the induction assumption: $\sum_{n=1}^{k} F_n^2 = F_k F_{k+1}$.

Add $F_{k+1}^2$ to both sides:

$$\sum_{n=1}^{k+1} F_n^2 = F_k F_{k+1} + F_{k+1}^2$$
$$= F_{k+1}(F_k + F_{k+1}) = F_{k+1}F_{k+2}.$$

139. Dozens of identities involving Fibonacci numbers have been developed. From the *Fibonacci Quarterly* alone we have gleaned the following: From Vol. 1 (1963), 13 identities on p. 66 of No. 1; 5 on p. 60, and 9 on p. 67 of No. 2. By means of a generalized program, John H. Halton has been able to derive 47 "generalized" identities which include "most of the identities, derivable as special cases ... which [he has] found in the literature, and a number of others." Vol. 3 (1965), p. 31.

# Index

# A CATALOG OF SELECTED DOVER
# BOOKS IN ALL FIELDS OF INTEREST

CONCERNING THE SPIRITUAL IN ART, Wassily Kandinsky. Pioneering work by father of abstract art. Thoughts on color theory, nature of art. Analysis of earlier masters. 12 illustrations. 80pp. of text. 5⅜ x 8½. 0-486-23411-8

CELTIC ART: The Methods of Construction, George Bain. Simple geometric techniques for making Celtic interlacements, spirals, Kells-type initials, animals, humans, etc. Over 500 illustrations. 160pp. 9 x 12. (Available in U.S. only.) 0-486-22923-8

AN ATLAS OF ANATOMY FOR ARTISTS, Fritz Schider. Most thorough reference work on art anatomy in the world. Hundreds of illustrations, including selections from works by Vesalius, Leonardo, Goya, Ingres, Michelangelo, others. 593 illustrations. 192pp. 7⅛ x 10¼. 0-486-20241-0

CELTIC HAND STROKE-BY-STROKE (Irish Half-Uncial from "The Book of Kells"): An Arthur Baker Calligraphy Manual, Arthur Baker. Complete guide to creating each letter of the alphabet in distinctive Celtic manner. Covers hand position, strokes, pens, inks, paper, more. Illustrated. 48pp. 8¼ x 11. 0-486-24336-2

EASY ORIGAMI, John Montroll. Charming collection of 32 projects (hat, cup, pelican, piano, swan, many more) specially designed for the novice origami hobbyist. Clearly illustrated easy-to-follow instructions insure that even beginning papercrafters will achieve successful results. 48pp. 8¼ x 11. 0-486-27298-2

BLOOMINGDALE'S ILLUSTRATED 1886 CATALOG: Fashions, Dry Goods and Housewares, Bloomingdale Brothers. Famed merchants' extremely rare catalog depicting about 1,700 products: clothing, housewares, firearms, dry goods, jewelry, more. Invaluable for dating, identifying vintage items. Also, copyright-free graphics for artists, designers. Co-published with Henry Ford Museum & Greenfield Village. 160pp. 8¼ x 11. 0-486-25780-0

THE ART OF WORLDLY WISDOM, Baltasar Gracian. "Think with the few and speak with the many," "Friends are a second existence," and "Be able to forget" are among this 1637 volume's 300 pithy maxims. A perfect source of mental and spiritual refreshment, it can be opened at random and appreciated either in brief or at length. 128pp. 5⅜ x 8½. 0-486-44034-6

JOHNSON'S DICTIONARY: A Modern Selection, Samuel Johnson (E. L. McAdam and George Milne, eds.). This modern version reduces the original 1755 edition's 2,300 pages of definitions and literary examples to a more manageable length, retaining the verbal pleasure and historical curiosity of the original. 480pp. 5³⁄₁₆ x 8¼. 0-486-44089-3

ADVENTURES OF HUCKLEBERRY FINN, Mark Twain, Illustrated by E. W. Kemble. A work of eternal richness and complexity, a source of ongoing critical debate, and a literary landmark, Twain's 1885 masterpiece about a barefoot boy's journey of self-discovery has enthralled readers around the world. This handsome clothbound reproduction of the first edition features all 174 of the original black-and-white illustrations. 368pp. 5⅜ x 8½. 0-486-44322-1

STICKLEY CRAFTSMAN FURNITURE CATALOGS, Gustav Stickley and L. & J. G. Stickley. Beautiful, functional furniture in two authentic catalogs from 1910. 594 illustrations, including 277 photos, show settles, rockers, armchairs, reclining chairs, bookcases, desks, tables. 183pp. 6½ x 9¼. 0-486-23838-5

AMERICAN LOCOMOTIVES IN HISTORIC PHOTOGRAPHS: 1858 to 1949, Ron Ziel (ed.). A rare collection of 126 meticulously detailed official photographs, called "builder portraits," of American locomotives that majestically chronicle the rise of steam locomotive power in America. Introduction. Detailed captions. xi+ 129pp. 9 x 12. 0-486-27393-8

AMERICA'S LIGHTHOUSES: An Illustrated History, Francis Ross Holland, Jr. Delightfully written, profusely illustrated fact-filled survey of over 200 American light-houses since 1716. History, anecdotes, technological advances, more. 240pp. 8 x 10¾. 0-486-25576-X

TOWARDS A NEW ARCHITECTURE, Le Corbusier. Pioneering manifesto by founder of "International School." Technical and aesthetic theories, views of industry, economics, relation of form to function, "mass-production split" and much more. Profusely illustrated. 320pp. 6⅛ x 9¼. (Available in U.S. only.) 0-486-25023-7

HOW THE OTHER HALF LIVES, Jacob Riis. Famous journalistic record, expos-ing poverty and degradation of New York slums around 1900, by major social reformer. 100 striking and influential photographs. 233pp. 10 x 7⅞. 0-486-22012-5

FRUIT KEY AND TWIG KEY TO TREES AND SHRUBS, William M. Harlow. One of the handiest and most widely used identification aids. Fruit key covers 120 deciduous and evergreen species; twig key 160 deciduous species. Easily used. Over 300 photographs. 126pp. 5⅜ x 8½. 0-486-20511-8

COMMON BIRD SONGS, Dr. Donald J. Borror. Songs of 60 most common U.S. birds: robins, sparrows, cardinals, bluejays, finches, more–arranged in order of increasing complexity. Up to 9 variations of songs of each species.
Cassette and manual 0-486-99911-4

ORCHIDS AS HOUSE PLANTS, Rebecca Tyson Northen. Grow cattleyas and many other kinds of orchids–in a window, in a case, or under artificial light. 63 illustrations. 148pp. 5⅜ x 8½. 0-486-23261-1

MONSTER MAZES, Dave Phillips. Masterful mazes at four levels of difficulty. Avoid deadly perils and evil creatures to find magical treasures. Solutions for all 32 exciting illustrated puzzles. 48pp. 8¼ x 11. 0-486-26005-4

MOZART'S DON GIOVANNI (DOVER OPERA LIBRETTO SERIES), Wolfgang Amadeus Mozart. Introduced and translated by Ellen H. Bleiler. Standard Italian libretto, with complete English translation. Convenient and thoroughly portable–an ideal companion for reading along with a recording or the performance itself. Introduction. List of characters. Plot summary. 121pp. 5¼ x 8½. 0-486-24944-1

FRANK LLOYD WRIGHT'S DANA HOUSE, Donald Hoffmann. Pictorial essay of residential masterpiece with over 160 interior and exterior photos, plans, eleva-tions, sketches and studies. 128pp. 9¼ x 10¾. 0-486-29120-0

THE CLARINET AND CLARINET PLAYING, David Pino. Lively, comprehensive work features suggestions about technique, musicianship, and musical interpretation, as well as guidelines for teaching, making your own reeds, and preparing for public performance. Includes an intriguing look at clarinet history. "A godsend," *The Clarinet,* Journal of the International Clarinet Society. Appendixes. 7 illus. 320pp. 5⅜ x 8½. 0-486-40270-3

HOLLYWOOD GLAMOR PORTRAITS, John Kobal (ed.). 145 photos from 1926-49. Harlow, Gable, Bogart, Bacall; 94 stars in all. Full background on photographers, technical aspects. 160pp. 8⅜ x 11¼. 0-486-23352-9

THE RAVEN AND OTHER FAVORITE POEMS, Edgar Allan Poe. Over 40 of the author's most memorable poems: "The Bells," "Ulalume," "Israfel," "To Helen," "The Conqueror Worm," "Eldorado," "Annabel Lee," many more. Alphabetic lists of titles and first lines. 64pp. 5³⁄₁₆ x 8¼. 0-486-26685-0

PERSONAL MEMOIRS OF U. S. GRANT, Ulysses Simpson Grant. Intelligent, deeply moving firsthand account of Civil War campaigns, considered by many the finest military memoirs ever written. Includes letters, historic photographs, maps and more. 528pp. 6⅛ x 9¼. 0-486-28587-1

POE ILLUSTRATED: Art by Doré, Dulac, Rackham and Others, selected and edited by Jeff A. Menges. More than 100 compelling illustrations, in brilliant color and crisp black-and-white, include scenes from "The Raven," "The Pit and the Pendulum," "The Gold-Bug," and other stories and poems. 96pp. 8⅜ x 11.
0-486-45746-X

RUSSIAN STORIES/RUSSKIE RASSKAZY: A Dual-Language Book, edited by Gleb Struve. Twelve tales by such masters as Chekhov, Tolstoy, Dostoevsky, Pushkin, others. Excellent word-for-word English translations on facing pages, plus teaching and study aids, Russian/English vocabulary, biographical/critical introductions, more. 416pp. 5⅜ x 8½. 0-486-26244-8

PHILADELPHIA THEN AND NOW: 60 Sites Photographed in the Past and Present, Kenneth Finkel and Susan Oyama. Rare photographs of City Hall, Logan Square, Independence Hall, Betsy Ross House, other landmarks juxtaposed with contemporary views. Captures changing face of historic city. Introduction. Captions. 128pp. 8¼ x 11. 0-486-25790-8

NORTH AMERICAN INDIAN LIFE: Customs and Traditions of 23 Tribes, Elsie Clews Parsons (ed.). 27 fictionalized essays by noted anthropologists examine religion, customs, government, additional facets of life among the Winnebago, Crow, Zuni, Eskimo, other tribes. 480pp. 6⅛ x 9¼. 0-486-27377-6

TECHNICAL MANUAL AND DICTIONARY OF CLASSICAL BALLET, Gail Grant. Defines, explains, comments on steps, movements, poses and concepts. 15-page pictorial section. Basic book for student, viewer. 127pp. 5⅜ x 8½.
0-486-21843-0

THE MALE AND FEMALE FIGURE IN MOTION: 60 Classic Photographic Sequences, Eadweard Muybridge. 60 true-action photographs of men and women walking, running, climbing, bending, turning, etc., reproduced from a rare 19th-century masterpiece. vi + 121pp. 9 x 12. 0-486-24745-7

ANIMALS: 1,419 Copyright-Free Illustrations of Mammals, Birds, Fish, Insects, etc., Jim Harter (ed.). Clear wood engravings present, in extremely lifelike poses, over 1,000 species of animals. One of the most extensive pictorial sourcebooks of its kind. Captions. Index. 284pp. 9 x 12. 0-486-23766-4

1001 QUESTIONS ANSWERED ABOUT THE SEASHORE, N. J. Berrill and Jacquelyn Berrill. Queries answered about dolphins, sea snails, sponges, starfish, fishes, shore birds, many others. Covers appearance, breeding, growth, feeding, much more. 305pp. 5¼ x 8¼. 0-486-23366-9

ATTRACTING BIRDS TO YOUR YARD, William J. Weber. Easy-to-follow guide offers advice on how to attract the greatest diversity of birds: birdhouses, feeders, water and waterers, much more. 96pp. 5³⁄₁₆ x 8¼. 0-486-28927-3

MEDICINAL AND OTHER USES OF NORTH AMERICAN PLANTS: A Historical Survey with Special Reference to the Eastern Indian Tribes, Charlotte Erichsen-Brown. Chronological historical citations document 500 years of usage of plants, trees, shrubs native to eastern Canada, northeastern U.S. Also complete identifying information. 343 illustrations. 544pp. 6½ x 9¼. 0-486-25951-X

STORYBOOK MAZES, Dave Phillips. 23 stories and mazes on two-page spreads: Wizard of Oz, Treasure Island, Robin Hood, etc. Solutions. 64pp. 8¼ x 11. 0-486-23628-5

AMERICAN NEGRO SONGS: 230 Folk Songs and Spirituals, Religious and Secular, John W. Work. This authoritative study traces the African influences of songs sung and played by black Americans at work, in church, and as entertainment. The author discusses the lyric significance of such songs as "Swing Low, Sweet Chariot," "John Henry," and others and offers the words and music for 230 songs. Bibliography. Index of Song Titles. 272pp. 6½ x 9¼. 0-486-40271-1

MOVIE-STAR PORTRAITS OF THE FORTIES, John Kobal (ed.). 163 glamor, studio photos of 106 stars of the 1940s: Rita Hayworth, Ava Gardner, Marlon Brando, Clark Gable, many more. 176pp. 8⅜ x 11¼. 0-486-23546-7

YEKL and THE IMPORTED BRIDEGROOM AND OTHER STORIES OF YIDDISH NEW YORK, Abraham Cahan. Film Hester Street based on *Yekl* (1896). Novel, other stories among first about Jewish immigrants on N.Y.'s East Side. 240pp. 5⅜ x 8½. 0-486-22427-9

SELECTED POEMS, Walt Whitman. Generous sampling from *Leaves of Grass.* Twenty-four poems include "I Hear America Singing," "Song of the Open Road," "I Sing the Body Electric," "When Lilacs Last in the Dooryard Bloom'd," "O Captain! My Captain!"—all reprinted from an authoritative edition. Lists of titles and first lines. 128pp. 5³⁄₁₆ x 8¼. 0-486-26878-0

SONGS OF EXPERIENCE: Facsimile Reproduction with 26 Plates in Full Color, William Blake. 26 full-color plates from a rare 1826 edition. Includes "The Tyger," "London," "Holy Thursday," and other poems. Printed text of poems. 48pp. 5¼ x 7. 0-486-24636-1

THE BEST TALES OF HOFFMANN, E. T. A. Hoffmann. 10 of Hoffmann's most important stories: "Nutcracker and the King of Mice," "The Golden Flowerpot," etc. 458pp. 5⅜ x 8½. 0-486-21793-0

THE BOOK OF TEA, Kakuzo Okakura. Minor classic of the Orient: entertaining, charming explanation, interpretation of traditional Japanese culture in terms of tea ceremony. 94pp. 5⅜ x 8½. 0-486-20070-1

FRENCH STORIES/CONTES FRANÇAIS: A Dual-Language Book, Wallace Fowlie. Ten stories by French masters, Voltaire to Camus: "Micromegas" by Voltaire; "The Atheist's Mass" by Balzac; "Minuet" by de Maupassant; "The Guest" by Camus, six more. Excellent English translations on facing pages. Also French-English vocabulary list, exercises, more. 352pp. 5⅜ x 8½.  0-486-26443-2

CHICAGO AT THE TURN OF THE CENTURY IN PHOTOGRAPHS: 122 Historic Views from the Collections of the Chicago Historical Society, Larry A. Viskochil. Rare large-format prints offer detailed views of City Hall, State Street, the Loop, Hull House, Union Station, many other landmarks, circa 1904-1913. Introduction. Captions. Maps. 144pp. 9⅜ x 12¼.  0-486-24656-6

OLD BROOKLYN IN EARLY PHOTOGRAPHS, 1865–1929, William Lee Younger. Luna Park, Gravesend race track, construction of Grand Army Plaza, moving of Hotel Brighton, etc. 157 previously unpublished photographs. 165pp. 8⅞ x 11¾.  0-486-23587-4

THE MYTHS OF THE NORTH AMERICAN INDIANS, Lewis Spence. Rich anthology of the myths and legends of the Algonquins, Iroquois, Pawnees and Sioux, prefaced by an extensive historical and ethnological commentary. 36 illustrations. 480pp. 5⅜ x 8½.  0-486-25967-6

AN ENCYCLOPEDIA OF BATTLES: Accounts of Over 1,560 Battles from 1479 B.C. to the Present, David Eggenberger. Essential details of every major battle in recorded history from the first battle of Megiddo in 1479 B.C. to Grenada in 1984. List of Battle Maps. New Appendix covering the years 1967–1984. Index. 99 illustrations. 544pp. 6½ x 9¼.  0-486-24913-1

SAILING ALONE AROUND THE WORLD, Captain Joshua Slocum. First man to sail around the world, alone, in small boat. One of the great feats of seamanship told in delightful manner. 67 illustrations. 294pp. 5⅜ x 8½.  0-486-20326-3

ANARCHISM AND OTHER ESSAYS, Emma Goldman. Powerful, penetrating, prophetic essays on direct action, role of minorities, prison reform, puritan hypocrisy, violence, etc. 271pp. 5⅜ x 8½.  0-486-22484-8

MYTHS OF THE HINDUS AND BUDDHISTS, Ananda K. Coomaraswamy and Sister Nivedita. Great stories of the epics; deeds of Krishna, Shiva, taken from puranas, Vedas, folk tales; etc. 32 illustrations. 400pp. 5⅜ x 8½.  0-486-21759-0

MY BONDAGE AND MY FREEDOM, Frederick Douglass. Born a slave, Douglass became outspoken force in antislavery movement. The best of Douglass' autobiographies. Graphic description of slave life. 464pp. 5⅜ x 8½.  0-486-22457-0

FOLLOWING THE EQUATOR: A Journey Around the World, Mark Twain. Fascinating humorous account of 1897 voyage to Hawaii, Australia, India, New Zealand, etc. Ironic, bemused reports on peoples, customs, climate, flora and fauna, politics, much more. 197 illustrations. 720pp. 5⅜ x 8½.  0-486-26113-1

GREAT SPEECHES BY AMERICAN WOMEN, edited by James Daley. Here are 21 legendary speeches from the country's most inspirational female voices, including Sojourner Truth, Susan B. Anthony, Eleanor Roosevelt, Hillary Rodham Clinton, Nancy Pelosi, and many others. 192pp. 5³⁄₁₆ x 8¼.  0-486-46141-6

THE MYTHS OF GREECE AND ROME, H. A. Guerber. A classic of mythology, generously illustrated, long prized for its simple, graphic, accurate retelling of the principal myths of Greece and Rome, and for its commentary on their origins and significance. With 64 illustrations by Michelangelo, Raphael, Titian, Rubens, Canova, Bernini and others. 480pp. 5⅜ x 8½.  0-486-27584-1

PSYCHOLOGY OF MUSIC, Carl E. Seashore. Classic work discusses music as a medium from psychological viewpoint. Clear treatment of physical acoustics, auditory apparatus, sound perception, development of musical skills, nature of musical feeling, host of other topics. 88 figures. 408pp. 5⅜ x 8½. 0-486-21851-1

LIFE IN ANCIENT EGYPT, Adolf Erman. Fullest, most thorough, detailed older account with much not in more recent books, domestic life, religion, magic, medicine, commerce, much more. Many illustrations reproduce tomb paintings, carvings, hieroglyphs, etc. 597pp. 5⅜ x 8½. 0-486-22632-8

SUNDIALS, Their Theory and Construction, Albert Waugh. Far and away the best, most thorough coverage of ideas, mathematics concerned, types, construction, adjusting anywhere. Simple, nontechnical treatment allows even children to build several of these dials. Over 100 illustrations. 230pp. 5⅜ x 8½. 0-486-22947-5

GREAT SPEECHES BY AFRICAN AMERICANS: Frederick Douglass, Sojourner Truth, Dr. Martin Luther King, Jr., Barack Obama, and Others, edited by James Daley. Tracing the struggle for freedom and civil rights across two centuries, this anthology comprises speeches by Martin Luther King, Jr., Marcus Garvey, Malcolm X, Barack Obama, and many other influential figures. 160pp. 5³⁄₁₆ x 8¼.
0-486-44761-8

OLD-TIME VIGNETTES IN FULL COLOR, Carol Belanger Grafton (ed.). Over 390 charming, often sentimental illustrations, selected from archives of Victorian graphics—pretty women posing, children playing, food, flowers, kittens and puppies, smiling cherubs, birds and butterflies, much more. All copyright-free. 48pp. 9¼ x 12¼.
0-486-27269-9

PERSPECTIVE FOR ARTISTS, Rex Vicat Cole. Depth, perspective of sky and sea, shadows, much more, not usually covered. 391 diagrams, 81 reproductions of drawings and paintings. 279pp. 5⅜ x 8½. 0-486-22487-2

DRAWING THE LIVING FIGURE, Joseph Sheppard. Innovative approach to artistic anatomy focuses on specifics of surface anatomy, rather than muscles and bones. Over 170 drawings of live models in front, back and side views, and in widely varying poses. Accompanying diagrams. 177 illustrations. Introduction. Index. 144pp. 8⅜ x11¼. 0-486-26723-7

GOTHIC AND OLD ENGLISH ALPHABETS: 100 Complete Fonts, Dan X. Solo. Add power, elegance to posters, signs, other graphics with 100 stunning copyright-free alphabets: Blackstone, Dolbey, Germania, 97 more—including many lower-case, numerals, punctuation marks. 104pp. 8⅛ x 11. 0-486-24695-7

THE BOOK OF WOOD CARVING, Charles Marshall Sayers. Finest book for beginners discusses fundamentals and offers 34 designs. "Absolutely first rate . . . well thought out and well executed."—E. J. Tangerman. 118pp. 7¾ x 10⅝. 0-486-23654-4

ILLUSTRATED CATALOG OF CIVIL WAR MILITARY GOODS: Union Army Weapons, Insignia, Uniform Accessories, and Other Equipment, Schuyler, Hartley, and Graham. Rare, profusely illustrated 1846 catalog includes Union Army uniform and dress regulations, arms and ammunition, coats, insignia, flags, swords, rifles, etc. 226 illustrations. 160pp. 9 x 12. 0-486-24939-5

WOMEN'S FASHIONS OF THE EARLY 1900s: An Unabridged Republication of "New York Fashions, 1909," National Cloak & Suit Co. Rare catalog of mail-order fashions documents women's and children's clothing styles shortly after the turn of the century. Captions offer full descriptions, prices. Invaluable resource for fashion, costume historians. Approximately 725 illustrations. 128pp. 8⅜ x 11¼. 0-486-27276-1

# CATALOG OF DOVER BOOKS

HOW TO DO BEADWORK, Mary White. Fundamental book on craft from simple projects to five-bead chains and woven works. 106 illustrations. 142pp. 5⅜ x 8.
0-486-20697-1

THE 1912 AND 1915 GUSTAV STICKLEY FURNITURE CATALOGS, Gustav Stickley. With over 200 detailed illustrations and descriptions, these two catalogs are essential reading and reference materials and identification guides for Stickley furniture. Captions cite materials, dimensions and prices. 112pp. 6½ x 9¼. 0-486-26676-1

SIX GREAT DIALOGUES: Apology, Crito, Phaedo, Phaedrus, Symposium, The Republic, Plato, translated by Benjamin Jowett. Plato's Dialogues rank among Western civilization's most important and influential philosophical works. These 6 selections of his major works explore a broad range of enduringly relevant issues. Authoritative Jowett translations. 480pp. 5³⁄₁₆ x 8¼.            0-486-45465-7

DEMONOLATRY: An Account of the Historical Practice of Witchcraft, Nicolas Remy, edited with an Introduction and Notes by Montague Summers, translated by E. A. Ashwin. This extremely influential 1595 study was frequently cited at witchcraft trials. In addition to lurid details of satanic pacts and sexual perversity, it presents the particulars of numerous court cases. 240pp. 6½ x 9¼.            0-486-46137-8

VICTORIAN FASHIONS AND COSTUMES FROM HARPER'S BAZAAR, 1867–1898, Stella Blum (ed.). Day costumes, evening wear, sports clothes, shoes, hats, other accessories in over 1,000 detailed engravings. 320pp. 9⅜ x 12¼.
0-486-22990-4

THE LONG ISLAND RAIL ROAD IN EARLY PHOTOGRAPHS, Ron Ziel. Over 220 rare photos, informative text document origin (1844) and development of rail service on Long Island. Vintage views of early trains, locomotives, stations, passengers, crews, much more. Captions. 8⅞ x 11¾.            0-486-26301-0

VOYAGE OF THE LIBERDADE, Joshua Slocum. Great 19th-century mariner's thrilling, first-hand account of the wreck of his ship off South America, the 35-foot boat he built from the wreckage, and its remarkable voyage home. 128pp. 5⅜ x 8½.
0-486-40022-0

TEN BOOKS ON ARCHITECTURE, Vitruvius. The most important book ever written on architecture. Early Roman aesthetics, technology, classical orders, site selection, all other aspects. Morgan translation. 331pp. 5⅜ x 8½.     0-486-20645-9

THE HUMAN FIGURE IN MOTION, Eadweard Muybridge. More than 4,500 stopped-action photos, in action series, showing undraped men, women, children jumping, lying down, throwing, sitting, wrestling, carrying, etc. 390pp. 7⅞ x 10⅝.
0-486-20204-6 Clothbd.

TREES OF THE EASTERN AND CENTRAL UNITED STATES AND CANADA, William M. Harlow. Best one-volume guide to 140 trees. Full descriptions, woodlore, range, etc. Over 600 illustrations. Handy size. 288pp. 4½ x 6⅜.       0-486-20395-6

MY FIRST BOOK OF TCHAIKOVSKY: Favorite Pieces in Easy Piano Arrangements, edited by David Dutkanicz. These special arrangements of favorite Tchaikovsky themes are ideal for beginner pianists, child or adult. Contents include themes from "The Nutcracker," "March Slav," Symphonies Nos. 5 and 6, "Swan Lake," "Sleeping Beauty," and more. 48pp. 8¼ x 11.            0-486-46416-4

BIG BOOK OF MAZES AND LABYRINTHS, Walter Shepherd. 50 mazes and labyrinths in all–classical, solid, ripple, and more–in one great volume. Perfect inexpensive puzzler for clever youngsters. Full solutions. 112pp. 8⅛ x 11. 0-486-22951-3

PIANO TUNING, J. Cree Fischer. Clearest, best book for beginner, amateur. Simple repairs, raising dropped notes, tuning by easy method of flattened fifths. No previous skills needed. 4 illustrations. 201pp. 5⅜ x 8½.            0-486-23267-0

HINTS TO SINGERS, Lillian Nordica. Selecting the right teacher, developing confidence, overcoming stage fright, and many other important skills receive thoughtful discussion in this indispensible guide, written by a world-famous diva of four decades' experience. 96pp. 5⅜ x 8½. 0-486-40094-8

THE COMPLETE NONSENSE OF EDWARD LEAR, Edward Lear. All nonsense limericks, zany alphabets, Owl and Pussycat, songs, nonsense botany, etc., illustrated by Lear. Total of 320pp. 5⅜ x 8½. (Available in U.S. only.) 0-486-20167-8

VICTORIAN PARLOUR POETRY: An Annotated Anthology, Michael R. Turner. 117 gems by Longfellow, Tennyson, Browning, many lesser-known poets. "The Village Blacksmith," "Curfew Must Not Ring Tonight," "Only a Baby Small," dozens more, often difficult to find elsewhere. Index of poets, titles, first lines. xxiii + 325pp. 5⅝ x 8¼. 0-486-27044-0

DUBLINERS, James Joyce. Fifteen stories offer vivid, tightly focused observations of the lives of Dublin's poorer classes. At least one, "The Dead," is considered a masterpiece. Reprinted complete and unabridged from standard edition. 160pp. 5³⁄₁₆ x 8¼. 0-486-26870-5

THE LITTLE RED SCHOOLHOUSE, Eric Sloane. Harkening back to a time when the three Rs stood for reading, 'riting, and religion, Sloane's sketchbook explores the history of early American schools. Includes marvelous illustrations of one-room New England schoolhouses, desks, and benches. 48pp. 8¼ x 11. 0-486-45604-8

THE BOOK OF THE SACRED MAGIC OF ABRAMELIN THE MAGE, translated by S. MacGregor Mathers. Medieval manuscript of ceremonial magic. Basic document in Aleister Crowley, Golden Dawn groups. 268pp. 5⅜ x 8½. 0-486-23211-5

THE BATTLES THAT CHANGED HISTORY, Fletcher Pratt. Eminent historian profiles 16 crucial conflicts, ancient to modern, that changed the course of civilization. 352pp. 5⅜ x 8½. 0-486-41129-X

NEW RUSSIAN-ENGLISH AND ENGLISH-RUSSIAN DICTIONARY, M. A. O'Brien. This is a remarkably handy Russian dictionary, containing a surprising amount of information, including over 70,000 entries. 366pp. 4½ x 6⅛. 0-486-20208-9

NEW YORK IN THE FORTIES, Andreas Feininger. 162 brilliant photographs by the well-known photographer, formerly with *Life* magazine. Commuters, shoppers, Times Square at night, much else from city at its peak. Captions by John von Hartz. 181pp. 9¼ x 10¾. 0-486-23585-8

INDIAN SIGN LANGUAGE, William Tomkins. Over 525 signs developed by Sioux and other tribes. Written instructions and diagrams. Also 290 pictographs. 111pp. 6⅛ x 9¼. 0-486-22029-X

ANATOMY: A Complete Guide for Artists, Joseph Sheppard. A master of figure drawing shows artists how to render human anatomy convincingly. Over 460 illustrations. 224pp. 8⅜ x 11¼. 0-486-27279-6

MEDIEVAL CALLIGRAPHY: Its History and Technique, Marc Drogin. Spirited history, comprehensive instruction manual covers 13 styles (ca. 4th century through 15th). Excellent photographs; directions for duplicating medieval techniques with modern tools. 224pp. 8⅜ x 11¼. 0-486-26142-5

DRIED FLOWERS: How to Prepare Them, Sarah Whitlock and Martha Rankin. Complete instructions on how to use silica gel, meal and borax, perlite aggregate, sand and borax, glycerine and water to create attractive permanent flower arrangements. 12 illustrations. 32pp. 5⅜ x 8½. 0-486-21802-3

EASY-TO-MAKE BIRD FEEDERS FOR WOODWORKERS, Scott D. Campbell. Detailed, simple-to-use guide for designing, constructing, caring for and using feeders. Text, illustrations for 12 classic and contemporary designs. 96pp. 5⅜ x 8½. 0-486-25847-5

THE COMPLETE BOOK OF BIRDHOUSE CONSTRUCTION FOR WOOD-WORKERS, Scott D. Campbell. Detailed instructions, illustrations, tables. Also data on bird habitat and instinct patterns. Bibliography. 3 tables. 63 illustrations in 15 figures. 48pp. 5¼ x 8½. 0-486-24407-5

SCOTTISH WONDER TALES FROM MYTH AND LEGEND, Donald A. Mackenzie. 16 lively tales tell of giants rumbling down mountainsides, of a magic wand that turns stone pillars into warriors, of gods and goddesses, evil hags, powerful forces and more. 240pp. 5⅜ x 8½. 0-486-29677-6

THE HISTORY OF UNDERCLOTHES, C. Willett Cunnington and Phyllis Cunnington. Fascinating, well-documented survey covering six centuries of English undergarments, enhanced with over 100 illustrations: 12th-century laced-up bodice, footed long drawers (1795), 19th-century bustles, l9th-century corsets for men, Victorian "bust improvers," much more. 272pp. 5⅜ x 8¼. 0-486-27124-2

FIRST FRENCH READER: A Beginner's Dual-Language Book, edited and translated by Stanley Appelbaum. This anthology introduces fifty legendary writers—Voltaire, Balzac, Baudelaire, Proust, more—through passages from The Red and the Black, Les Misérables, Madame Bovary, and other classics. Original French text plus English translation on facing pages. 240pp. 5⅜ x 8½. 0-486-46178-5

WILBUR AND ORVILLE: A Biography of the Wright Brothers, Fred Howard. Definitive, crisply written study tells the full story of the brothers' lives and work. A vividly written biography, unparalleled in scope and color, that also captures the spirit of an extraordinary era. 560pp. 6⅛ x 9¼. 0-486-40297-5

THE ARTS OF THE SAILOR: Knotting, Splicing and Ropework, Hervey Garrett Smith. Indispensable shipboard reference covers tools, basic knots and useful hitches; handsewing and canvas work, more. Over 100 illustrations. Delightful reading for sea lovers. 256pp. 5⅜ x 8½. 0-486-26440-8

FRANK LLOYD WRIGHT'S FALLINGWATER: The House and Its History, Second, Revised Edition, Donald Hoffmann. A total revision—both in text and illustrations—of the standard document on Fallingwater, the boldest, most personal architectural statement of Wright's mature years, updated with valuable new material from the recently opened Frank Lloyd Wright Archives. "Fascinating"—*The New York Times*. 116 illustrations. 128pp. 9¼ x 10¾. 0-486-27430-6

PHOTOGRAPHIC SKETCHBOOK OF THE CIVIL WAR, Alexander Gardner. 100 photos taken on field during the Civil War. Famous shots of Manassas Harper's Ferry, Lincoln, Richmond, slave pens, etc. 244pp. 10⅝ x 8¼. 0-486-22731-6

FIVE ACRES AND INDEPENDENCE, Maurice G. Kains. Great back-to-the-land classic explains basics of self-sufficient farming. The one book to get. 95 illustrations. 397pp. 5⅜ x 8½. 0-486-20974-1

CATALOG OF DOVER BOOKS

A MODERN HERBAL, Margaret Grieve. Much the fullest, most exact, most useful compilation of herbal material. Gigantic alphabetical encyclopedia, from aconite to zedoary, gives botanical information, medical properties, folklore, economic uses, much else. Indispensable to serious reader. 161 illustrations. 888pp. 6½ x 9¼. 2-vol. set. (Available in U.S. only.)          Vol. I: 0-486-22798-7     Vol. II: 0-486-22799-5

HIDDEN TREASURE MAZE BOOK, Dave Phillips. Solve 34 challenging mazes accompanied by heroic tales of adventure. Evil dragons, people-eating plants, blood-thirsty giants, many more dangerous adversaries lurk at every twist and turn. 34 mazes, stories, solutions. 48pp. 8¼ x 11.                                    0-486-24566-7

LETTERS OF W. A. MOZART, Wolfgang A. Mozart. Remarkable letters show bawdy wit, humor, imagination, musical insights, contemporary musical world; includes some letters from Leopold Mozart. 276pp. 5⅜ x 8½.          0-486-22859-2

BASIC PRINCIPLES OF CLASSICAL BALLET, Agrippina Vaganova. Great Russian theoretician, teacher explains methods for teaching classical ballet. 118 illustrations. 175pp. 5⅜ x 8½.                                          0-486-22036-2

THE JUMPING FROG, Mark Twain. Revenge edition. The original story of The Celebrated Jumping Frog of Calaveras County, a hapless French translation, and Twain's hilarious "retranslation" from the French. 12 illustrations. 66pp. 5⅜ x 8½.
                                                                        0-486-22686-7

BEST REMEMBERED POEMS, Martin Gardner (ed.). The 126 poems in this superb collection of 19th- and 20th-century British and American verse range from Shelley's "To a Skylark" to the impassioned "Renascence" of Edna St. Vincent Millay and to Edward Lear's whimsical "The Owl and the Pussycat." 224pp. 5⅜ x 8½.
                                                                        0-486-27165-X

COMPLETE SONNETS, William Shakespeare. Over 150 exquisite poems deal with love, friendship, the tyranny of time, beauty's evanescence, death and other themes in language of remarkable power, precision and beauty. Glossary of archaic terms. 80pp. 5³⁄₁₆ x 8¼.                                          0-486-26686-9

HISTORIC HOMES OF THE AMERICAN PRESIDENTS, Second, Revised Edition, Irvin Haas. A traveler's guide to American Presidential homes, most open to the public, depicting and describing homes occupied by every American President from George Washington to George Bush. With visiting hours, admission charges, travel routes. 175 photographs. Index. 160pp. 8¼ x 11.          0-486-26751-2

THE WIT AND HUMOR OF OSCAR WILDE, Alvin Redman (ed.). More than 1,000 ripostes, paradoxes, wisecracks: Work is the curse of the drinking classes; I can resist everything except temptation; etc. 258pp. 5⅜ x 8½.          0-486-20602-5

SHAKESPEARE LEXICON AND QUOTATION DICTIONARY, Alexander Schmidt. Full definitions, locations, shades of meaning in every word in plays and poems. More than 50,000 exact quotations. 1,485pp. 6½ x 9¼. 2-vol. set.
                              Vol. 1: 0-486-22726-X     Vol. 2: 0-486-22727-8

SELECTED POEMS, Emily Dickinson. Over 100 best-known, best-loved poems by one of America's foremost poets, reprinted from authoritative early editions. No comparable edition at this price. Index of first lines. 64pp. 5³⁄₁₆ x 8¼. 0-486-26466-1

THE INSIDIOUS DR. FU-MANCHU, Sax Rohmer. The first of the popular mystery series introduces a pair of English detectives to their archnemesis, the diabolical Dr. Fu-Manchu. Flavorful atmosphere, fast-paced action, and colorful characters enliven this classic of the genre. 208pp. 5³⁄₁₆ x 8¼.          0-486-29898-1

THE MALLEUS MALEFICARUM OF KRAMER AND SPRENGER, translated by Montague Summers. Full text of most important witchhunter's "bible," used by both Catholics and Protestants. 278pp. 6⅝ x 10. 0-486-22802-9

SPANISH STORIES/CUENTOS ESPAÑOLES: A Dual-Language Book, Angel Flores (ed.). Unique format offers 13 great stories in Spanish by Cervantes, Borges, others. Faithful English translations on facing pages. 352pp. 5⅜ x 8½.
0-486-25399-6

GARDEN CITY, LONG ISLAND, IN EARLY PHOTOGRAPHS, 1869–1919, Mildred H. Smith. Handsome treasury of 118 vintage pictures, accompanied by carefully researched captions, document the Garden City Hotel fire (1899), the Vanderbilt Cup Race (1908), the first airmail flight departing from the Nassau Boulevard Aerodrome (1911), and much more. 96pp. 8⅞ x 11¾. 0-486-40669-5

OLD QUEENS, N.Y., IN EARLY PHOTOGRAPHS, Vincent F. Seyfried and William Asadorian. Over 160 rare photographs of Maspeth, Jamaica, Jackson Heights, and other areas. Vintage views of DeWitt Clinton mansion, 1939 World's Fair and more. Captions. 192pp. 8⅞ x 11. 0-486-26358-4

CAPTURED BY THE INDIANS: 15 Firsthand Accounts, 1750-1870, Frederick Drimmer. Astounding true historical accounts of grisly torture, bloody conflicts, relentless pursuits, miraculous escapes and more, by people who lived to tell the tale. 384pp. 5⅜ x 8½. 0-486-24901-8

THE WORLD'S GREAT SPEECHES (Fourth Enlarged Edition), Lewis Copeland, Lawrence W. Lamm, and Stephen J. McKenna. Nearly 300 speeches provide public speakers with a wealth of updated quotes and inspiration–from Pericles' funeral oration and William Jennings Bryan's "Cross of Gold Speech" to Malcolm X's powerful words on the Black Revolution and Earl of Spenser's tribute to his sister, Diana, Princess of Wales. 944pp. 5⅜ x 8⅜. 0-486-40903-1

THE BOOK OF THE SWORD, Sir Richard F. Burton. Great Victorian scholar/adventurer's eloquent, erudite history of the "queen of weapons"–from prehistory to early Roman Empire. Evolution and development of early swords, variations (sabre, broadsword, cutlass, scimitar, etc.), much more. 336pp. 6⅛ x 9¼.
0-486-25434-8

AUTOBIOGRAPHY: The Story of My Experiments with Truth, Mohandas K. Gandhi. Boyhood, legal studies, purification, the growth of the Satyagraha (nonviolent protest) movement. Critical, inspiring work of the man responsible for the freedom of India. 480pp. 5⅜ x 8½. (Available in U.S. only.) 0-486-24593-4

CELTIC MYTHS AND LEGENDS, T. W. Rolleston. Masterful retelling of Irish and Welsh stories and tales. Cuchulain, King Arthur, Deirdre, the Grail, many more. First paperback edition. 58 full-page illustrations. 512pp. 5⅜ x 8½. 0-486-26507-2

THE PRINCIPLES OF PSYCHOLOGY, William James. Famous long course complete, unabridged. Stream of thought, time perception, memory, experimental methods; great work decades ahead of its time. 94 figures. 1,391pp. 5⅜ x 8½. 2-vol. set.
Vol. I: 0-486-20381-6      Vol. II: 0-486-20382-4

THE WORLD AS WILL AND REPRESENTATION, Arthur Schopenhauer. Definitive English translation of Schopenhauer's life work, correcting more than 1,000 errors, omissions in earlier translations. Translated by E. F. J. Payne. Total of 1,269pp. 5⅜ x 8½. 2-vol. set.     Vol. 1: 0-486-21761-2     Vol. 2: 0-486-21762-0

LIGHT AND SHADE: A Classic Approach to Three-Dimensional Drawing, Mrs. Mary P. Merrifield. Handy reference clearly demonstrates principles of light and shade by revealing effects of common daylight, sunshine, and candle or artificial light on geometrical solids. 13 plates. 64pp. 5⅜ x 8½. 0-486-44143-1

ASTROLOGY AND ASTRONOMY: A Pictorial Archive of Signs and Symbols, Ernst and Johanna Lehner. Treasure trove of stories, lore, and myth, accompanied by more than 300 rare illustrations of planets, the Milky Way, signs of the zodiac, comets, meteors, and other astronomical phenomena. 192pp. 8⅜ x 11.
0-486-43981-X

JEWELRY MAKING: Techniques for Metal, Tim McCreight. Easy-to-follow instructions and carefully executed illustrations describe tools and techniques, use of gems and enamels, wire inlay, casting, and other topics. 72 line illustrations and diagrams. 176pp. 8¼ x 10⅞. 0-486-44043-5

MAKING BIRDHOUSES: Easy and Advanced Projects, Gladstone Califf. Easy-to-follow instructions include diagrams for everything from a one-room house for bluebirds to a forty-two-room structure for purple martins. 56 plates; 4 figures. 80pp. 8¾ x 6⅜. 0-486-44183-0

LITTLE BOOK OF LOG CABINS: How to Build and Furnish Them, William S. Wicks. Handy how-to manual, with instructions and illustrations for building cabins in the Adirondack style, fireplaces, stairways, furniture, beamed ceilings, and more. 102 line drawings. 96pp. 8¾ x 6⅜. 0-486-44259-4

THE SEASONS OF AMERICA PAST, Eric Sloane. From "sugaring time" and strawberry picking to Indian summer and fall harvest, a whole year's activities described in charming prose and enhanced with 79 of the author's own illustrations. 160pp. 8¼ x 11. 0-486-44220-9

THE METROPOLIS OF TOMORROW, Hugh Ferriss. Generous, prophetic vision of the metropolis of the future, as perceived in 1929. Powerful illustrations of towering structures, wide avenues, and rooftop parks—all features in many of today's modern cities. 59 illustrations. 144pp. 8¼ x 11. 0-486-43727-2

THE PATH TO ROME, Hilaire Belloc. This 1902 memoir abounds in lively vignettes from a vanished time, recounting a pilgrimage on foot across the Alps and Apennines in order to "see all Europe which the Christian Faith has saved." 77 of the author's original line drawings complement his sparkling prose. 272pp. 5⅜ x 8½.
0-486-44001-X

THE HISTORY OF RASSELAS: Prince of Abissinia, Samuel Johnson. Distinguished English writer attacks eighteenth-century optimism and man's unrealistic estimates of what life has to offer. 112pp. 5⅜ x 8½. 0-486-44094-X

A VOYAGE TO ARCTURUS, David Lindsay. A brilliant flight of pure fancy, where wild creatures crowd the fantastic landscape and demented torturers dominate victims with their bizarre mental powers. 272pp. 5⅜ x 8½. 0-486-44198-9

Paperbound unless otherwise indicated. Available at your book dealer, online at **www.doverpublications.com**, or by writing to Dept. GI, Dover Publications, Inc., 31 East 2nd Street, Mineola, NY 11501. For current price information or for free catalogs (please indicate field of interest), write to Dover Publications or log on to **www.doverpublications.com** and see every Dover book in print. Dover publishes more than 400 books each year on science, elementary and advanced mathematics, biology, music, art, literary history, social sciences, and other areas.